Lean TPM
A Blueprint for Change

Lean TPM
A Blueprint for Change

Second edition

Dennis McCarthy & Nick Rich

AMSTERDAM • BOSTON • HEIDELBERG • LONDON
NEW YORK • OXFORD • PARIS • SAN DIEGO
SAN FRANCISCO • SINGAPORE • SYDNEY • TOKYO
Butterworth-Heinemann is an imprint of Elsevier

Butterworth-Heinemann is an imprint of Elsevier
The Boulevard, Langford Lane, Kidlington, Oxford OX5 1GB, UK
225 Wyman Street, Waltham MA 02451, USA

Library of Congress Cataloging-in-Publication Data
A catalog record for this book is available from the Library of Congress

British Library Cataloguing-in-Publication Data
A catalogue record for this book is available from the British Library

ISBN: 978-0-08-100090-8

For Information on all Butterworth-Heinemann publications
visit our website at http://store.elsevier.com/

Working together
to grow libraries in
developing countries

www.elsevier.com • www.bookaid.org

Acquisition Editor: Hayley Gray
Editorial Project Manager: Cari Owen
Production Project Manager: Jason Mitchell
Designer: Maria Inês Cruz

Typeset by TNQ Books and Journals
www.tnq.co.in

Printed and bound in the United Kingdom

Contents

Preface

The purpose of this book is to set out how the convergence of Lean Thinking and Total Productive Maintenance (TPM) presents a comprehensive blueprint for business-led change. This also sets out how leadership and strategic thinking are an important part of the recipe for successful and sustained improvement. This is not intended as a comprehensive guide to Lean Thinking or TPM techniques, but as a user manual on how to deliver business benefits from their application. It is assumed that the reader already has some awareness of the basics of these world-class manufacturing techniques. The book is organised to be read from cover to cover.

Acknowledgements

My motivation for writing the first version of this book almost 10 years ago was to explore why Lean and TPM improvement tools worked well in some businesses and not in others. The answer only really became clear after the last chapter had been written and the book was in the hands of the publisher. The answer starting from the chapters of Lean TPM was collective leadership. Organisations that knew, in detail, where they wanted to get to, made the fastest progress. Good progress was also made by organisations that were not sure about what they wanted but knew what they had was not good enough. The weakest results came when organisations were content with what they had. This included companies with a few individuals who saw the need to do more but without the collective will to move beyond the current level of performance. These are symptoms of a lack of collective leadership. The term collective leadership refers to the fact that the delivery of improved performance involves leaders at all levels in the organisation (senior managers, middle/front line managers and shop-floor teams). In successful organisations, these leaders develop and deploy winning growth strategies in a way that engages their peers and direct reports with a purpose to believe in. Unfortunately, despite the plethora of texts on improvement tools, there is a dearth of 'how to' guides on how to use these to develop winning strategies and engage the workforce with their delivery.

Since then I have been on a mission to capture and share the practical ways in which organisations learn how to lead their customer agenda and engage the workforce behind a single improvement agenda. This has not only confirmed the value of the Lean TPM approach of the earlier book but has added more clarity about where the synergy comes from. This includes the integration of the TPM **change master plan**, a powerful and practical transformational guide with **value stream mapping** and Lean **policy deployment**. Enhancing the TPM **bottom-up stepwise process** with its clear incremental goals and accountabilities to encompass CANDO/5S and Lean standard work. Adapting the **stepwise audit** process as a framework for coaching and team-based recognition to reward and reinforce behavioural change. A process which also supports the development of high performance teamwork capabilities. Extending the concept of **improving effectiveness** through small group activity to include all business processes and combining that with a **systems thinking** approach to improve process resilience. Using **early management** to develop and the design of operations that support **flow, flexibility and focused improvement**.

In developing the material for this book I have worked with many inspirational managers. 3M Aycliffe, BP, Post Danmark, General Motors, Johnson Matthey, Aunt Bessie foods, SPSL Carmeuse, Heineken, Newsprinters.

Sorry guys! but too many to mention but you know who you are. I owe you for letting me work with your teams. I have also learned enthusiastically from those prepared to share the lessons of their improvement journey. I particularly appreciate the opportunity to work with such an experienced practitioner of Lean as Nick Rich. This has significantly smoothed the process of blending together two sets of experience to produce a robust and practical improvement tool. Last but not least, thanks go to Karen, my wife, who not only had to suffer intense bouts of writing 'the book' for almost 2 years but has now had to suffer the update.

Dennis McCarthy

It is an interesting time in history where recession has bitten deep and this has prompted lots of experimentation and innovation as businesses weather the storm. Many businesses have returned to TPM to get the most from their production systems and strangely enough many service industries (including healthcare processes) have begun to study and implement the principles that underpin this fundamental basis of 'world class' performance. Many of these new and established TPM advocates have been kind enough to send us their stories and successes – so I would like to say a big 'thank you' to these individuals and organizations.

I would really like to thank my close family, Fiona, Dan and Josh. Josh probably more than most as our days were cut short to get back to 'the book'. Sorry!

The Royal Mint always deserves a mention for me. They are a constant source of innovation and I have been lucky to have been involved with the Mint for many decades now! I have even been an employee when I joined the 2012 Olympic and Paralympic Medal Production programme for the 2012 London games. 4700 perfect medals, produced on time and record breakers in their own right as the biggest and heaviest medals ever produced. I learned a lot from that team and my long-term friendships with the Directors, Change Champions, Team Leaders and specialists. Of particular mention are Adam Lawrence, Phil Carpenter, Leighton John, Gavin Elliott, Steve Gregory, Tony Baker, Gwyn Roberts, Ashley Gibbs, Simon Janczewski, Kevin Jones, Kev Chamberlain, Mike 'Doc' Jones, Bethan Parry, Lyn 'Turbo', James Attridge, Paul Binning, Phil Samuel, Peter Murphy (now at Marstons), John Bunney, Glen Evans, Louise Terry, Ann Jessop, Ian Jones, Ross Davies, Bev Thomas, Marie Buckley, Sian Merry, Steve Tilley, Big Steve Davies, James Thomas, Marc Hanson, Matt James, Graham Smith, Chris Williams, Fergus Feaney, the supply chain team and many more. And not forgetting Dennis Scott – now retired and former mentor of mine – Merlin in a lab and a really lovely man with a brain like a planet!

Other individuals and groups who deserve a thank you include the John Bicheno, Dan Jones, Professor Matthew Cooke, Dr Ann Esain, Dr Sharon Williams, Dr Pauline Found, Dr Maneesh Kumar, John Darlington, Dr Peter Treadwell, my colleagues at Swansea University, Warwick Medical School, Buckingham University, Cardiff Business School and at Cardiff School of Management.

The Welsh Blood Service, English Blood Service, Prism Medical, my mad scientist friends in Oxford, and NMI, all deserve special mention and so too my former students who now manage some great organizations from aerospace, to bottling wine, food, white goods, nuclear fuel, automotive and many more. Finally my PhD students past and present – a constant source of pride for me and individuals who always ask 'why?' just to keep me on my toes!! Thank you all!

Nick Rich

List of Abbreviations

A3	A process method for structured problem-solving
ABC	Classification approach (criticality)
ABCD Goals	Accidents, Breakdowns, Contamination and Defect Analysis
CANDO	Cleaning, arranging, neatening, discipline and order in workplace organisation (aka 5S)
CBM	Condition-based monitoring
CFM	Cross-functional Management
D2D	Door to door OEE measurement
EEM	Early Equipment Management
EM	Early Management
EPM	Early Product Management
F2F	Floor to floor OEE measurement
FLM	Front Line Management
JIPM	Japanese Institute of Plant Maintenance
JIT	Just in time
NVA	Non-value-adding activity
OEE	Overall Equipment Effectiveness
OTIF	On time in full customer deliveries
PD	Policy Deployment aka Hoshin Kanri
PPM	Parts per million defects
QC	Quality control
QDC also QCD	Quality, delivery and cost performance objectives
QFD	Quality Function Deployment design approach
R&D	Research and development
RCM	Reliability Centred Maintenance
S2C	Supply chain OEE measurement
SMED	Single Minute Exchange of Dies (Quick Changeover of machinery)
SPL	Single Point Lesson Instruction document
TCO	Total cost of ownership
TPM	Total Productive Maintenance/Total Productive Manufacturing
TPM5	Fifth Bi-annual European TPM Forum
TPS	Toyota Production System
TQM	Total quality management
TRAC	Team review and coaching
VOC	Voice of the customer performance expectations
VSM	Value stream mapping analysis
WCM	World Class Manufacturing

Chapter | One

The Business of Survival and Growth

Let us face facts – most manufacturing businesses are under pressure to compete and to extract greater profit from what they convert. The ability to do so is determined by good product design (that minimises materials and maximises operational efficiency through good design for manufacture/assembly) and waste-free production (where every second of every shift is used to produce perfect quality outputs).

At the heart of the competition is the need to survive, to grow, to capture the benchmark position for their industry and also to improve efficiency and effectiveness at a rate faster than the competition (wherever they may be located). Lean total productive manufacturing (TPM) provides the solid foundations for a world class production system by developing the 3Rs of reliable processes, robust value streams and resilient organisations that can survive and prosper in today's marketplace. This may sound quite a Darwinian statement – and it is – only the fittest and fitting (fitting profitably with market requirements) will survive. A great product will never achieve its potential if the production system that delivers it is unreliable and a great production system can never turn a bad product into a good one! These are 'certains' in an uncertain world. Even the great quality guru Deming was heard, on many occasions, to state that 'survival is optional!'

To thrive, businesses need the strategic foundations of good product flow, reduced variation and process flexibility to offer the quality, order fulfilment lead time and flexibility to compete in modern markets with the minimum of finished goods stock. Achieving this is not possible without the application of both Lean and TPM principles and techniques as equal partners.

If an organisation has the highest quality, the shortest 'turnaround and delivery times' for products and the smallest of stocks, one of the few events that can stop you from delivering a customer's requirements is equipment or process failure. These conditions, of minimal customer stocks, and a desire for eight deliveries per day for the vehicle assembly factories of Toyota in Japan set the challenge and started the process that has led to the Lean TPM model.

Lean TPM is a major competitive weapon that supports low-cost production and a market advantage based on cost competitiveness and it also supports

1

a strategy of product and service differentiation by generating an equipment capability where processes are reliable, changed over in minimal time and where it is possible to customise products to order. The power of a manufacturing process comes from a partnership between operational and sales staff. Sales staff do not always know best – in fact some sales staff know a lot about the product but very little about how it is made and many operational staff know a lot about how to manufacture the product but not its application or how customers derive value from it. A strategy therefore provides focus and allows these two vital and mutually dependent activities to come together – to sell solutions to customer requirements profitably. Such a strategy is well beyond the mere application of a few Lean tools and point improvements to limited areas of the business – it is all embracing and focused on providing value by transforming business performance.

However, historical studies show that many manufacturers have fundamental weaknesses in their competitive weaponry. Some businesses do not have a formal business strategy or a manufacturing strategy that outlines how manufacturing will allow marketing staff to win orders and many businesses struggle to focus and manage change effectively (Brown, 1996). These processes are further complicated because markets and competitors do not stand still and what used to win orders last year is considered just the basic level of performance needed to begin to negotiate with a customer – a qualifier to do business if you like (Hill, 1985). Market requirements are constantly changing and adding more uncertainty and confusion for managers. Those who do nothing and simply react to markets will have an uncertain future. They will be in crisis management and try to sell anything to anyone. Competition is an inevitable part of manufacturing today and the ability of a firm to compete is the final arbiter of the longevity of any business. A great product, a good brand name and capable employees are not enough to guarantee survival – the management of good thinking, good people, good products and outstanding processes is the answer. Management means the approach to leading change, the methods used to engage with the workforce's creativity and a system of systematic change to deploy the right techniques to constantly improve. We will return to this theme of policy deployment later in the book and the role of TPM as a solid foundation upon which to make quality and delivery promises to customers that maximise cash flow and customer service.

1.1 THE NEW COMPETITIVE CONDITIONS

The modern competitive conditions have generated a new 'set of rules' for manufacturers and the transparency of the Internet means customers really do hold the power in any relationship. The new customer/consumer rules include the provision of the highest level of customer service, the delivery of quality products in shorter lead times and product proliferation to offer variety to customers (Brown, 1996). If you take a few minutes to consider what life was like 10 years ago and compare it to now, your business has probably moved on substantially.

In the past, businesses recorded product quality in terms of percentage defects produced during manufacturing but this measure changed quickly to that of 'parts per million' levels; businesses typically offer more products and service combinations than before and will have halved their lead times for new and existing products. Taking a few more minutes, you may like to contemplate the future and guess what? These performance indicators are likely to get tougher and tougher. The new rules of competition demand the effective management of the rate of change within the business and the elimination of all unnecessary waste or costs in order to provide the ultimate levels of customer service throughout the firm.

Reliable processes are mandatory in today's competitive world. In business-to-business relations, it is typical to find minimal levels of stock and frequent deliveries – this is not possible without TPM. Good-quality systems and a very fast logistics system cannot overcome catastrophic and frequent equipment and process failures. It can be no surprise that suppliers to Toyota developed the TPM system to overcome the challenge of a just-in-time logistics system which demanded eight deliveries per day to resupply only 2 h of stock which was held at the line side of the vehicle assembler (and not in a big warehouse close by). Personalisation and a great brand reputation for products and services can only be achieved with a very flexible process – another objective of TPM. Indeed today we often regard Dell as the benchmark – and a business that turned an industry on its head through the disruptive power of a very flexible and customer-focused production facility and supporting supply chain.

Whilst many businesses were content to build large batches of stock and sell it from warehouses, Dell assembled customised laptops to order in very short lead times so that Internet-based customers get the exact product they want, pay at the time they configure their computer and then receive it within a matter of hours. Such a high-performance production system did not come about by chance – it was the result of a customer-driven strategy that ignored the conventional way of thinking and just applying a few management techniques here and there. The Dell operating model truly reset customer expectations and provided a benchmark for the industry.

Lean TPM is a vehicle to deliver sustained 'world class' performance because the feedback generated from progress through TPM master plan milestones combined with the Lean operations model provides a road map to the next generation of products, services and equipment. Robust, reliable and resilient systems result from Lean TPM and they offer the certainty that management decisions (to reduce stocks or to work to a customer pull) will be realised. Without Lean TPM, a production and service system is incomplete. If **quality** improvement programmes and great product designs determine the quality of the product and process and the **delivery** of a system is determined by the use of Lean flow and pull systems then the **cost** of operating the system (without high and unnecessary stocks, without large production batches and without inflexible processes) is determined by the quality teams and the Lean flow teams (ideally, this is one team that has matured rather than different groups) and the critical knowledge they gain from TPM. TPM effectively hardwires a system. Without TPM, great and effective customer service is a temporary phenomenon. It will not last. TPM, often misrepresented as operator maintenance, will not last unless its master plan road map is aligned with the Lean principles of enhancing customer value and seeking perfection.

External pressures

The race to compete and to survive is a difficult one. Markets are full of pressures that increase uncertainty for managers. In today's markets, managers must be able to sense and respond to critical changes if they are to make the right and timely improvement activities within the factory. These pressures come from a variety of sources including the government (laws and taxation), customers who expect ever-improving levels of service, consumer groups who inevitably seek to lower prices, competitors, parent corporations and shareholders who demand a return for their investment.

The new competitive conditions are far removed from those of the past and challenge strategies such that we can no longer assume that:
- past business success is a guarantee of future survival;
- product patents will protect a manufacturer from competition;
- buying the latest technology will provide a means of defence against competition.

Technology or products by themselves are not enough to guarantee survival. The countries of the developing world are eager to take their place in the world economy. They know that they have the opportunity to leapfrog the traditional costly batch and queue approach in favour of a more efficient low-inventory, high-flow and high-quality operations. With the support of organisations looking for low-cost supplies they have also developed the management skills and expertise to run their operations at high levels of effectiveness.

Many offshore competitors also have the advantage of governments prepared to offer advantageous tax allowances to attract inward investment. These advantages should not be overstated. They have their problems including poor infrastructure, low home demand, poor material supply and, in some cases, corruption. So developing countries do not have it all their own way but when these constraints are lifted, they will be even more competitive. Naturally, as this happens their costs will also rise. The size of their domestic markets and therefore global consumption will increase as a result. These pressures for change (Table 1.1), shaping the shifting sands of the future market, can be predicted without the aid of a crystal ball. The only uncertainty is when!

All the pressures and opportunities, from outside and from within, mount up to a major challenge. However, the challenge is surmountable if management can:
- harness the intellectual capability of the complete workforce;
- target the creativity on making better products, more cheaply;
- achieve 'world class' manufacturing standards that set your organisation apart from the competition.

TABLE 1.1 Pressures for Change

1. New and emerging manufacturing economies with low labour costs are attracted to mature Western markets where they can exploit their 'cost advantage'.
2. The power of the Internet in purchasing materials and components on a global scale and therefore access to alternative suppliers has increased exponentially. As such, power has shifted to the customer/consumer.
3. Deregulation of world markets has resulted from international trading agreements and this has liberated trade and increased competition for manufacturers.
4. Corporations have the ability to switch production.
5. Pressure groups and lobbyists seeking to lower prices or convince the manufacturer to improve their performance in areas such as environmental management.
6. Shareholders who expect a 'year-on-year' improvement in the returns on their money invested and constantly compare these returns with what their money could earn elsewhere.
7. Customers expect product variety, continuously improving quality levels, lead time reduction and want their stocks reduced.
8. Customers that use your product in multiple of their products (such as the automotive industry) want to be assured that a product failure cannot happen and will not lead to product recalls from the consumer.

1.2 SILVER BULLETS, INITIATIVE FATIGUE AND FASHIONABLE MANAGEMENT

Management books and academic publications over the past 20 years have heralded new business models that would, if applied correctly, radically transform the firm into a 'world beater' capable of meeting the demands of the market and fending off competitors (Suzaki, 1987; Womack & Jones, 1996; Rother, 2010; Schmenner, 2012). The trend to find new models started in the 1980s when Japanese texts explaining certain manufacturing techniques of high-performing firms were translated into English and this created an interest in applying these techniques in non-Japanese workplaces – hopefully with a rise in performance. Many of these models were, however, to prove seductively rational but without a methodology for implementation and often no explanation as to why the technique was invented and what problem it solved. The techniques provided a methodology but offered no real advice concerning how to integrate them into business-wide improvement activities or what support activities would be necessary to ensure the techniques would 'stick'.

Managers in many countries readily adopted these new innovations. These managers were motivated by a number of reasons; some were keen to implement and be seen to be at the 'cutting edge', whilst others grasped at these new practices as if they were lifelines and implemented change with an air of desperation hoping for any form of improvement. Whilst some improvements were implemented and sustained many were not and ended in failure. Such failures did little to increase the credibility of managers with employees, trade unions and even customers. Today most authors are keen to explain their master plans, what their methods will solve and the human side of the interventions so that managers can maximise the chances of a successful and sustainable change (Mann, 2010; Liker & Convis, 2012).

1.3 WHY PROGRAMMES FAIL?

It is important to understand the causes of failure. Some improvement initiatives fail because they were applied in a piecemeal way – grafted onto existing practices and were rejected by those people who did not select or necessarily understand 'why' change was necessary but were tasked with implementing it and working in a new way (lack of a communicated master plan, promotion to all staff and its deployment as policy). Others fail because they were little more than 'technical quick fixes' and 'sticking plaster' solutions and many such changes were quickly reversed, failed or left as the manager jumped quickly into the next 'fix' (short-term thinking without a methodology for sustainable improvement). Examples of these programmes include attempts to compress the set-up time of machines in a belief that the company will be capable of producing high levels of variety and missing the fact that the existing machinery was not capable of meeting the quality tolerances needed of it. Or that the process of quick changeover was overreliant on a highly motivated team of employees – a team as motivated as the pit crews in a Formula One or NASCAR race. So whilst many have claimed massive theoretical 'time savings', often little is added to bottom-line profitability from such one-off 'point improvement activities'. The 'time savings' simply vanish because the related planning, quality, forecasting and inventory management functions do not change their policies or ways of working. As a result, the potential quality, inventory, lead time and productivity gains are not made. Production teams became bored (lack of sustainability and embedding the change) and a 'hard to reverse' fatigue sets in.

Single 'point improvement activities' fail because they are not supported by processes to standardise work and systematically improve workplace discipline, problem-solving, visual management and high-performance teamwork. In addition, they tend to be carried out in response to a specific issue rather than to raise end-to-end capabilities.

Without an effective production system and without a clear transformational goal – such as 'zero breakdowns' – it is impossible to create the pressure for change to challenge weak practices and sustain the learning cycle needed to adopt new patterns of behaviour. The development of a learning organisation is vital if the improvement cycle is to improve understanding and benefit from actions to simultaneously drive cost down whilst enhancing value to customers, shareholders and employees. Isolated and piecemeal change just does not work no matter how charismatic the change leader is! It simply is not sustainable.

The result is disappointment and a waste of management and employee time – we are all 'time poor' and we are rarely short of something else to do with our time. The desire to improve is part of our natural instinct but there is a profound behavioural difference between organisations where actions are triggered to **avoid** issues, problems and those that lead their organisations **towards** a desired state. Organisations that deliver lasting improvement adopt the latter approach using a change master plan containing clear transitions to systematically release the full potential of the end-to-end operation.

Developing a master plan as a road map **towards** a desired state is a key part of the management Continuous Improvement (CI) role. This is the process by which managers clarify strategic intent, identify capability gaps, deploy policies to main the creative pressure for flawless execution and enhance strategic competence. The outputs provide a map to guide the transitions from reactive to preventive to step-out performance improvement. The challenge for management is to 'think through' what the business needs to succeed, plot the forward journey and engage all levels of the organisation with that. Business school case studies are littered with examples where what 'glittered' and promised so much at the beginning delivered very little, damaged the credibility of managers as business leaders and became a barrier to future initiatives. Despite that fact that the road map to exemplar performance is well documented, less than 1% of organisations achieve it. Why is that?

Management outlook

It may seem obvious but the collective management outlook is one of the key drivers of long-term business performance yet it is a topic ignored by many improvement tools. Three prevalent themes within behavioural economics provide an insight into why the behaviours of managers can sometimes defy logic. Firstly, people often make decisions based on rules of thumb rather than logic. That is not all bad because the use of experience is essential to guide the strategic direction. Unfortunately, despite the evidence to the contrary, those that believe in the success of simplistic mechanistic cost down protocols are hardwired to repeat past failures.

Secondly, decisions can be flawed because they are made within a frame of reference which is too narrow. For example, some taxi drivers have a daily target in mind for the money they want to make in a day and will stay at the wheel until that is made. This makes sense until you take a wider frame of reference. For example, on sunny days, when people are more likely to walk rather than take a cab, cabbies could end up working very late to achieve their target. When it is raining, when people want to take cabs, they may be able to achieve their target easily. With this wider frame of reference they may prefer to take the day off when it is sunny and work all day when it rains. Framing can be useful when blanking out trivial issues to concentrate on the important few. It can help to speed up decision-making in complex situations. However, it is important to know what you do not know and to be aware of how a frame of reference can influence decision-making. Even where managers have experience of successful improvement processes, replicating success in other organisations will require more than the transfer of tools and techniques. That is why training alone seldom delivers value.

Finally, people tend to be overoptimistic about the future and conservative about change. Again, this behavioural pattern can be useful when navigating through uncertain times. Maintaining a clear strategic direction is an essential part of the recipe for long-term business success. Unfortunately, this behaviour can also result in group thinking where organisations can become dominated by patterns of behaviour that worked in the past but that do not meet the challenges of the modern marketplace.

A classic response to a declining financial position in organisations (without an aligned long-term management outlook) is to attempt to 'weather the storm'. This is the 'denial' phase. When results get so bad that they cannot be ignored, a new commercial model is sought. This avoidance behaviour frequently results in redundancy (downsizing) and a new management structure. The new management structure often sees the middle management team restructured with significant loss of knowledge assets (people) in the process (Figure 1.1).

Strategically, downsizing makes organisations vulnerable to take over and reduces the chances of long-term business survival. At the bottom of the trough, this is the path well trodden but research shows that it frequently does not work. Professors James Guthrie, from the University of Kansas, and Deepak Datta, from the University of Texas at Arlington, examined data on 122 firms that had engaged in downsizing and statistically analysed whether the programme had improved their profitability. And the answer was a plain and simple 'no'.

Exemplar organisations avoid the dilemma by pursuing a growth strategy and adopting behaviours and processes which progress **towards** a desired future state. This maintains a creative continuous improvement tension which counters the risk of complacency. It encourages proactive leadership, 'humbleness', engagement and collaborative patterns of behaviour which harness the collective potential of their organisation.

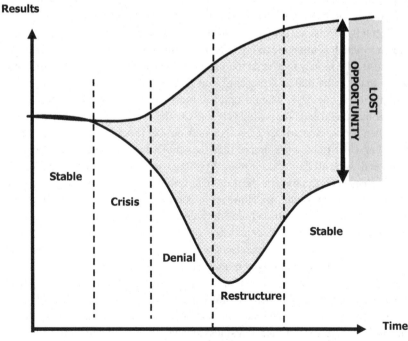

FIGURE 1.1 Crisis management stages: patterns of lost opportunity.

The potential value of adopting a proactive company-wide management outlook compared to that of a 'point improvement' management outlook is impossible to measure but the route map to exemplar performance resides with the former rather than the latter outlook.

At a strategic level, the behaviours of organisations that adopt an **avoid/**'point improvement outlook' are characterised by matching business activity to the state of current markets. This is no easy option for operations management. It is painful, and there are dozens of examples as to how this approach has resulted in 'death by a thousand cuts' as the course of the organisation is dictated by the chaos of market forces. The end game of this strategy is totally predictable. The organisation eventually falls below critical mass where the form of the business is unable to support the overheads necessary to operate in its chosen market. Merger/acquisition becomes the only alternative to receivership. As over 50% of these fail to produce real benefits, the cycle repeats itself.

Consider the following joke levelled at management by the workforce. An operations manager, on taking up a new position, found three sealed envelopes in the top drawer of his desk. The first envelope contains a page of paper that holds the following advice: 'Blame your predecessors and if things don't improve open envelope 2'. This works for a while but eventually the manager opens envelope 2 seeking advice. Envelope 2 contains a similar piece of paper that urges the management to 'Restructure the organisation and if this fails then open envelope 3'. Dutifully the manager follows the advice but eventually he finds he has to open the third envelope. This letter states 'Find another job but before leaving photocopy these three letters and leave them in the top drawer of this desk for your successor'.

As with all jokes, it has a ring of truth about it and presents a view of management, which is unable to manage, and can only react to events. This depressing reality occurs all too frequently due to:
- a lack of strategic clarity and analyses of market trends;
- inappropriate key performance measures;
- fragile technology;
- departmental/silo thinking.

At a strategic level, the behaviours of organisations that adopt a **towards/** proactive improvement outlook are characterised by activities to satisfy the four prerequisites for change. Each must be present if change is to be successful and sustaining. If one element is missing, then the model fails to work – the four prerequisites are:
- an understood **pressure** for the organisation to change;
- a clear, shared and reasoned/documented **vision** of the future that is communicated to everyone in the organisation;
- the deliberate creation of the **capacity** to engage in change including the knowledge and skills to do so;
- clear and well-defined **actionable first steps** in starting the process of change.

There are four prerequisites for change management – where all four elements are present then change will be supported but when just one element is missing

TABLE 1.2 The Four Prerequisites for Transformational Change

Your Changes Are	Pressure for Change	Vision	Capacity	Actionable First Steps
Successful	Yes	Yes	Yes	Yes
Lacks urgency and buy-in	No	Yes	Yes	Yes
Appears random and without logic	Yes	No	Yes	Yes
Frustrating – no one has time or skills	Yes	Yes	No	Yes
More thinking than doing – stagnation	Yes	Yes	Yes	No

then the change process is limited or unsuccessful. These four prerequisites are core to the Lean TPM process which is one of collaboration, inclusion of all stakeholders and based on an aligned middle management team. To be aligned, these key managers need to satisfy the four prerequisites of transformation (Table 1.2). As can be seen in Table 1.2, missing even one prerequisite will reduce the level of engagement of their direct reports. As discussed below, engagement personnel behind a common improvement agenda is key to learning and making the right investments in the softer side of Lean TPM.

The importance of engagement

Top-down-driven initiatives ignore the potential of the vast majority of problem-solvers in the firm. By creating a programme where 90% of employees receive rather than initiate improvements, management creates a dependency culture. The outcome is that individuals wait for others to change things. This is a learned condition where team members believe that they do not have the power to make change or would not be listened to if they do offer suggestions. In this environment, people get used to finding their fulfilment outside of work.

Naturally not every employee can be involved in every decision but employees are the experts in their individual areas of the transformation process and are also often customers. Not engaging all with the process of change ignores a major lever to accelerate the pace of improvement. In addition, this also increases the risk of failure by contributing to the following common reasons for the failure of improvement initiatives:

- the key 'influencer' left or moved positions;
- goals too distant or too vague to engage all levels of personnel;
- benefits/results disputed or not properly recorded;
- insufficient training;
- competing crisis distracts attention.

Scratching the surface of these responses using 'five why's' analysis suggests a common root cause for all these problems (see Table 1.3).

TABLE 1.3 Why the Programme Failed

The Problem: The programme failed when the champion left
1. **Why?** because those left behind were not motivated enough to press on with the changes needed.
2. **Why?** because the reasons for doing so were not compelling enough.
3. **Why?** because dealing with the barriers was more painful than living with the inefficiency.
4. **Why?** because there was not a collective will to change.
5. **Why?** because not enough people shared the belief change was really necessary. **Significant root cause: lack of clear business growth strategy or business leadership alignment or both.**

This is a pattern which is so common as to be familiar to most who have worked in medium and large organisations. The most common cause of CI failure can be summarised as one of weak management outlook.

The consequences of this deficiency is a failure of management to satisfy the four principles of employee engagement. These are:

1. A purpose to believe in (how can I contribute to the success of the organisation)
2. Support to adapt (what do I need to change)
3. Reinforcement and recognition (standardisation and praise when I get it right)
4. Consistent management role models (all managers preaching from the same hymn sheet)

When these principles are met, success is achieved where leaders help teams to find their own solutions to correctly defined priorities. Top-down direction is blended with engaged and questioning staff who discuss problems openly. Success is assured by leaders who communicate, communicate, communicate to help all levels and functions collaborate to learn how best to raise standards and improve the business and production system.

Cultural anchors

Management outlook and employee engagement define working relationships and reinforce the patterns of behaviour that shape organisational culture. Culture, or 'the way we do things around here', is driven by instinctive patterns of behaviour. We instinctively **avoid** discomfort/pain and move **towards** what we perceive as pleasurable. Many managers reading this book will begin to feel uncomfortable at this stage but please persevere, as an understanding of the wiring up behind **avoid** and **towards** behavioural triggers is critical to the design of initiatives that stick and sustain improvements. For those who want to really get into how people perceive change then reading the quirky book *Who Moved My Cheese?* (1999) by Dr Spencer Johnson is a good start – one of the lessons of the book is we repeat patterns of behaviours that are redundant.

To understand the 'wiring' that lies behind behaviours and how to influence culture it is important to recognise that:

1. Your 'gut instinct', when deciding what to do next, will typically be based on patterns of past experience. It will be formed based on your frame of reference at the time and you will favour things that worked for you in the past.

All employees, including managers, will have a natural tendency to avoid difficult problems – they are painful.

2. When making choices, your mind gives equal weight to ideas based on beliefs/rules of thumb and those based on fact. If there is a belief that 'They will never let us do that', it will not normally be subject to any test of logic.

These are two of the reasons why seemingly intelligent people sometimes do stupid things or do things that are satisfactory rather than optimal.

These emotional, instinctive patterns of behaviour can be positive. They help us to drive safely along the motorway at speed or to develop the co-ordination to 'bend it like Beckham'. The patterns can also be limiting and because they are instinctive, difficult to unlearn and replace. That is why making change happen can feel like walking in treacle.

The most powerful lever for change available to managers is the conversations they have with their direct reports and the questions they ask. To be effective, however, the conversations and questions need to be consistent. Not just at an individual manager level. The collective outlook of managers needs to be consistent so that proactive behaviours are rewarded and reinforced at every opportunity. Without this consistency, people can become confused about the management intent and will revert to type.

Why is culture important? Most manufacturers, speeding up the flow process of getting materials from 'door-to-door', especially with established operations would recognise the importance of people engagement (Liker & Hoseus, 2008). Whilst certain practices can be implemented by key 'technical staff functions' like condition-based maintenance (engineering) and advanced six sigma techniques (quality assurance), delivering lasting gains from such techniques requires front line 'buy-in' to adopt new ways of working and develop new skills (Womack & Jones, 1996; Liker & Convis, 2012). Even managers with newly established operations, 'mass employee' cultural development is needed to secure competitive advantage through 'zero loss' manufacturing. Culture also impacts on the pace of change and the ability to gain consensus amongst different groups.

So what can be done to tackle this important dimension of the continuous improvement agenda?

As explained earlier, the most powerful lever for change available to managers are the conversations they have with their direct reports and the questions they ask. That is because the questions they ask either challenge or reinforce the status quo through relatively simple cultural anchors illustrated in Figure 1.2.

As can be seen, only two of these anchors are within the scope of most improvement tools (organisational structure/internal systems and procedures). To deliver lasting change, pressure must be applied to all six anchors.

Where patterns of behaviour are limiting, organisations that are in dire, financial circumstances can adopt the 'burning platform' (avoid trigger) to deliver change but this only works whilst the fire is burning. The pace of change will slow down as the pain is reduced. As with a visit to the dentist to deal with a toothache, once the pain has gone, it takes different action triggers to change your dental hygiene to prevent future decay. To bring this closer

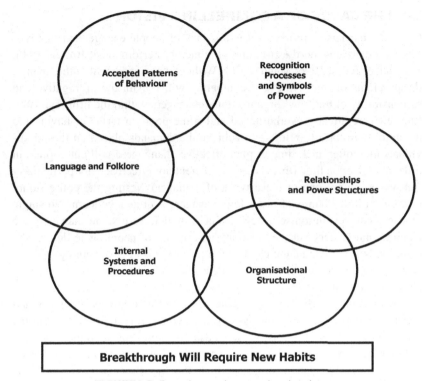

FIGURE 1.2 Restraints to change: cultural anchors.

to home, if the daily management conversations are mostly concerned with punishing the guilty, this is analogous to the visit to the dentist to deal with a toothache. Problems will be ignored/hidden, and the reality will only become visible when it can no longer be avoided. This is a symptom of cultures driven by pain. To develop a proactive culture, we need to establish a basis for daily management conversations that meets the principle of engagement, providing employees with a purpose to believe in and supporting their development, recognising and reinforcing good behaviours and providing consistent role models. This is where successful companies score over their less successful competitors. Although the difference between success and failure is far narrower than many would imagine, this will take more than a few well-chosen motivational slogans (and slogans are not conducive to business survival and prosperity – this is 1 of 14 points of great management that was highlighted by the influential quality guru W Edwards Deming). Most people instinctively know when they are being manipulated or when they are being sold something that feels false.

A consistent unifying vision must therefore be backed up with a systematic process to raise standards, increase flexibility and secure outstanding performance. Only then will organisation commit the resources needed to develop the collective competence to anticipate and lead the customer agenda (an outcome to which all world class organisations aspire).

1.4 THE VALUE OF A COMPELLING VISION

Of all the barriers to progress, it is this lack of people engagement that represents the greatest hurdles to improving factory performance (Brown, 1996; Womack & Jones, 1996; Mann, 2010). Without the engagement of the 'many', the problems of improving and competing will remain the prerogative and responsibility of the 'few' organisational managers within the firm. Typically 'the few' are already overburdened and 'time poor' in terms of having too much to do in too little time – to achieve the demands placed on these individuals they often lack time to plan effective change and will cut corners in order to deliver on time (this is neither efficient nor effective – nor particularly fun when you are one of these types of manager). Vesting the entire future of the firm in the hands of just a few managers is quite a concern. No single manager can revolutionise a new business model. Sure some managers are inspirational leaders but they also attract a group of people who deliver the changes. No single manager can hope to perform his or her planning role and also execute the huge amount of change needed to be competitive and stay ahead of the game.

By engaging workers, those people who determine factory efficiency and generate a stable material flow system, managers can be released to plan the future of the firm and to devise the key business improvement programmes that will transform business performance. An empowering vision and focus provide meaning for everyone in the organisation. In this way, they can identify with the challenge and become a part of it. If you think that this is unattainable, the next time you visit an exemplar company, observe how passionate the people are about their company. Organisations who find it difficult to achieve performance improvements seem to continuously change their business structures creating pain and uncertainty (in terms of roles and fit) for everyone.

So an appreciation of how best to engage and develop the skills of the workforce to increase the quality and quantity of improvement programmes within the door-to-door flow is highly important in the short term and also as the foundation for a sustainable improvement process.

Understanding motivation

The litmus test of a compelling vision is that it is capable of meeting the first principles of engagement: a purpose to believe in. That in turn can only be achieved when each worker finds some fulfilment in the process of change. Change brings uncertainty and the need to invest time in learning new patterns of working. That takes effort and as we have an inbred preference for the status quo, there must be something worthwhile in the change process. The alternative is for people willing to be dragged kicking and screaming into the new reality. On the other hand, if the change is of interest, it will be a case of 'light the blue touch paper and stand well clear'. So what motivates people?

Despite the fact that research into performance-related pay demonstrates its failure to sustain increased performance, there are still those who make the mistake of trying to buy co-operation. Just look at the banking system if you want to see the problems that such behaviour generates. If this stalwart of management tradition does not work, what does?

Research shows that in addition to our basic survival needs we are also driven to satisfy other needs such as identity, excitement, learning and to feel valued. These needs are met in many different ways but it is what drives football fans to follow their teams across the world, backpackers to endure the discomfort/personal risk and pilgrims to search for enlightenment. They have all found a 'compelling vision' that gives them focus, meaning and positive engagement (Figure 1.3).

What is in play, in most cases, is the pleasure gained by sharing experiences with others. That is, the compelling vision is as much an outcome of relationships as it is of individual endeavour. By creating the situation where people can work in teams to achieve a common goal, organisations can tap into a powerful motivator. One that provides individuals with certainty, variety, opportunities to learn, share ideas, connect with others and gain respect from their peers. Experience shows that improvement works best when it is a team-based rather than individual activity.

Working in teams

Although less exotic, the 'human need' to be part of a group and to generate good rapport by positive relationships and team development is another integral part of the improvement journey. A vehicle is to engage the organisation with competing in a 'future-focused' way. In practical terms, this means building the improvement process around balanced teams of five to seven people with a clear goal (aligned with the vision) and the skills/capability to achieve it.

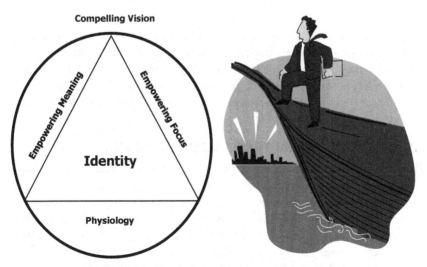

FIGURE 1.3 Developing a vision to engage individuals.

Unfortunately, although teamwork has been a feature of world class manufacturing (WCM) for over 20 years, scant attention is paid to the structured development process (Womack et al., 1990). The development of foundation-level team working does not mean sending everyone on 'outward bound' courses, although this may be useful in some circumstances, it is all about nurturing the most basic unit of the business – the controllers of tasks, processes and systems.

The following three-step process sets out a proven route for establishing basic-level team working:

- Step 1: Provide information to help employees to position themselves in the big picture. For example, raise an understanding of the entire supply chain of the firm and how quickening the material-to-cash cycle will improve the performance of the firm.
- Step 2: Define boundaries/focus/roles for each team and involve them in identifying improvement proposals that will speed up flow. To achieve this level of continuous flow, all barriers must be broken down in a logical sequence that starts with the availability and quality of each stage in the flow process.
- Step 3: Support the drive to achieve and sustain zero targets across the value chain from product design, through operations, final shipment and after-sales service of the product.

Refer back to Figure 1.2 and consider what impact this process could have on the cultural anchors in your organisation. Teams should control distinct 'whole tasks' that have distinct beginnings and handover points and teams must subscribe to the belief that every team has an internal (or external) customer and supplier. It is vitally important that they focus on their immediate customer and what the customer and they themselves need in terms of process flow.

The outcome is the creation of cells/teams who know how to develop proactive relationships to support the application of technical skills. This is one of the outcomes that can be delivered as part of a properly structured 5S/CANDO workplace organisation process (discussed in more detail in Chapter 5) or leader standard work for management grades. Once such basic team working capability has been mastered, it provides the foundation for the evolution of these teams to higher levels of empowerment and self-management (the complete High-Performance Teamwork Development process is set out in Chapter 3).

1.5 LEADING THE IMPROVEMENT PROCESS

The value of an organisation that practises a vision-led set of management behaviours is clear. In such an environment, teams can (with support) develop their ability to become self-managed. Leaders at all levels are able to delegate routine tasks to those who are nearest to the day-to-day operation and thereby create the necessary space and time for managers to direct and co-ordinate improvement efforts.

The senior manager leadership role concerns the 12 month + long term to strategic decision horizon. For middle managers, it is the medium- to long-term decision horizon and for front-line managers the short- to medium-term decision time horizon. There are a number of 'rules of thumb' for how much time each management level should devote to improvement and innovation activities and how much should be spent on the daily operations of the business. The division of time in this way underpins the current debate on Lean leadership and how much time should be engaged in Leader Standard Work – looking after daily management tours and reviews so that this time is standardised to allow the manager more time to engage in improvement activities. For a discussion of Leader Standard Work, see Mann (2010). The basic rule of thumb that was passed to Nick Rich on his first visit to study Toyota in Japan was that team members will spend 80%+ of their time adding value by working to set standards and up to 20% engaged in improvement activities. Middle managers spend approximately 40% of their time in daily standardised reviews and meetings to check that the production system is working and then 60% of their time in teams conducting system improvement programmes and senior directors would spend 20% of their time conducting their daily checks and 80% thinking about the future, assessing markets, products and competitors, reviewing new innovations and generally ensuring that the future competitiveness of the business is being developed (setting and explaining the vision and why it is important).

Identifying the problems facing manufacturing at each decision horizon is a relatively easy task. More difficult is the process of designing the improvement programme to engage the workforce and secure incremental mastery of key business processes (and by default the techniques needed). This can be likened to dialling a telephone number; the digits have to be selected in the right sequence to get the right connection. The paragraphs below set out an overview of the Lean TPM leadership journey.

Review/formalise current practices

The first stage in the commercial–cultural improvement model is to formalise current practices to standardise and simplify practices across the organisation. This removes recurring problems and provides a foundation for sharing ideas. As the initial layer of waste and inefficiency is removed, the root causes of management firefighting are revealed and meaningful pockets of management and specialist time can be released. This means that activities can be redefined and routine activities delegated. Learning takes place to release valuable specialist resources to address the next layer of waste/value. When these are brought under control, processes become simplified and in turn support further delegation.

Build rapport across functions and organisations

Cross-functional projects encourage a process of 'horizontal empowerment' to blur the traditional boundaries between functions and levels. Properly managed, it builds rapport and better working relationships which in turn improves communication. It also reduces the well-documented constraints to growth caused by the lack of availability of skilled labour. Talent management and personnel development are as important to future success as securing funding. Those organisations that have experience of Lean production and have engaged in autonomous maintenance will know well these issues and the benefits of correctly aligning the worker, learning and improvement activities.

The potential of these rapport-building skills extends beyond the boundaries of the organisation to include supplier relationships. Traditionally, they have been treated as 'enemies' accounting for a high percentage of manufacturing costs. This ignores the potential of supplier relationships built on releasing innovation in preference for one built on 'annual price negotiations'. This is despite the evidence that no matter how tight the legal contract, they are no substitute for a proactive working relationship.

So in many respects, the external world beyond the factory gate and modern competitive environment is more complicated than it could be because many firms have not found a methodology and business model with which to integrate and focus resources within the firm.

Learn from experience

The new 'manufacturing challenge' is therefore for managers to engage all levels of employees in building robust, dependable and flexible manufacturing processes that creates a purpose-designed manufacturing system of 'delay-free' material flow within and beyond the factory gate. The methodologies behind the technical, silver bullet tools and techniques are well covered and can achieve 'point benefits' for elements of the door-to-door flow but enlisting the 'hearts and minds' of an empowered workforce is the key to long-term and sustainable organisational learning that delivers results.

Encourage questions

Our desire to learn, with careful design, can be tapped into and used to support the pace of change of today's markets for manufactured products. At the heart of this learning capability lies understanding the customer and differentiating between what adds value and helps material to flow and what adds costs and waste. Teaching employees to understand and reflect upon how best to change the organisation is important to nurturing the 'learning culture'. It is no surprise that managers adopting this approach do not get upset when subordinates question their change programmes and decisions. Questions are one of the most important countermeasures to interrupt patterns of behaviour and counter 'group think'. They are to be encouraged. Those of you who have taken part in behavioural safety will recognise that questions are an important indicator of interest, learning and engagement.

Anticipate the growth stages

A recent study of the progress made by organisations on their journey to excellence shows that the learning mechanism is a key lever to developing collective

TABLE 1.4 Managing the Improvement Journey

Milestone	Improvement Landscape	Learning Goal
1	Set standards for equipment, processes and behaviours. Formalise current practices against them. Provide information and education to address recurring problems. Establish basic flow processes, remove excess inventory and control sources of dirt, dust and accelerated deterioration.	How to define and maintain basic conditions. Stabilise lead times. Establish basic team working practices. Engage all employees with the company competitive agenda.
2	Simplify and refine core activities and remove sources of accelerated deterioration to deliver the 'zero breakdown' goal. Refine flow layout and inventory/planning parameters to reflect true demand profile.	How to achieve 'zero breakdowns'. Compress internal lead times. Remove cross-functional barriers to secure high-performance teamwork.
3	Refocus specialist resources to optimise value-adding processes and address the causes of quality variations. Extend time between interventions (reel-to-reel running). Improve end-to-end supply chain performance to release step-out product and service capability. Improve project management of products, services and capital projects to achieve flawless operation from day 1 production.	How to achieve 'zero unplanned intervention'. Install low-cost automation. Establish flawless introduction of new products and processes. Remove cross-organisational barriers to compress total supply lead times and increase the pace of innovation.
4	Condition way of working to sustain process optimisation activities. Focus on step-out products and services to set the customer agenda. Reset the forward master plan based to ratchet up customer value and sustain step-out performance.	How to achieve 'no touch production'. Reduce new product time to market. Increased customer loyalty and growth.

capability as a means of accessing new and more productive patterns of working. The 'learning to improve' journey completed by exemplar organisations can be described as four learning landscapes as set out in Table 1.4. This is presented as a map of the leadership landscape rather than a fixed universal solution. The programme should be adapted to support the business drivers for the relevant industry sector, market segment and customer service strategy. This task of developing a relevant improvement master plan is discussed in more detail in Chapter 2.

Establish a single change agenda

A sustainable and continuously improving manufacturing system therefore can only be built incrementally. Each stage is mastered and a competence gained before moving to the next. This means mastering a basic level of competence in Process Stabilisation (see Chapter 5) before mastering Process Optimisation (see Chapter 6). It is important to understand the sequence and logic of

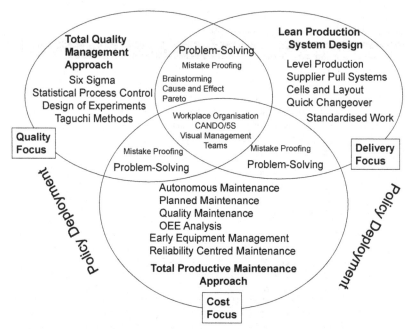

FIGURE 1.4 World class manufacturing techniques. Rich (2002).

production system development so that the learning/improvement process is linked to the delivery of increasing levels of manufacturing advantage. In turn, this becomes part of the mechanism to secure a compelling vision for the organisation (Figure 1.4).

Although 'zero accidents' and 'zero breakdowns' are the prime targets during the initial stages of stabilisation, actions to secure this will also increase operations effectiveness and efficiency. Assuming that there are no major safety hot spots, when selecting the area to start priority should be given to opportunities to improve the quality/competitive advantage of the entire linked chain of 'door-to-door' activities (both office and production area) to achieve a more competitive/higher added value offering to the customer (Table 1.5).

From here, the focus moves onto shortening the time from receipt of order to customer delivery and then to find new ways of offering high variety performance (in shorter and smaller lot sizes). This logic is quite easy to understand and each stage removes a level of waste and cost from the business. Obviously some investments will need to be made along the way but overall the important task of providing customer service should result in the total cost of manufacturing falling and productivity (including learning about how best to be productive) rising.

So, despite the complexity of the business problems facing manufacturing management, the logic of WCM performance is not that difficult to understand or build (Schonberger, 1986). In fact, clarity and consistency of purpose are essential levers in the process of securing lasting change. In this way, all

TABLE 1.5 The Five Sources of Manufacturing-Led Competitive Advantage

Competitive (Market) Advantage	Role of Business and Operations Management
Quality	To make things right based upon the needs of the customer
Speed	To make things fast and in shorter amounts of time
Delivery	To make things on time with dependable delivery times
Flexibility	To offer product variety and have the ability to change/update the product catalogue (what is made)
Cost	To make things at the lowest cost, to enhance margins within the guideline that the product is 'fit for purpose' and meets customer needs

Adapted from Slack (1991).

employees from senior managers, trade union officials, to the newest of operators and office workers can understand it. This form of incremental mastery allows each new chapter in the development of the company improvement process to be evolved with the workforce before each change theme is launched. In this manner, the management prerogative to lead change is reinforced and the application of techniques has a logic and a purpose rather than just asking employees to change without an understanding of the bigger picture. As most managers know from bitter experience this latter form of change management breeds frustration and often ends in disappointment.

1.6 LEAN TPM

So far in this chapter we have set out the challenges facing manufacturing and suggested that these can be met if management can:
- harness the intellectual capability of the complete workforce;
- target this creativity on making better products more cheaply;
- achieve WCM standards that set the organisation apart from the competition.

These are the goals of Lean TPM which are delivered by a process which combines the development of leadership and management best practice to secure long-lasting change. The key features and benefits of the approach are set out in the paragraphs below.

An approach based on proven business models

The combination of Lean thinking (Womack & Jones, 1996) and Total Productive Manufacturing (Lean TPM) applies the proven business models of 'world class' manufacturing firms who have learned how to dictate the rate of change and competitiveness of their chosen markets. This incorporates a change process which is far removed from 'blank silver bullets' and whilst each company must develop their own unique approach, the model provides many proven design principles for managers and the workforce.

Challenges limiting behaviours

Lean TPM creates a potent blend of case studies, principles and techniques to challenge old patterns of behaviour and replace them with a more versatile, flexible and proactive outlook.

Techniques to develop a compelling vision

The Lean TPM approach also presents a 'future state' business model within which empowerment and learning combine to allow 'mastery' of those key customer winning criteria that mark out high performers from the 'also-rans'. Both approaches promote the hypothesis that the future of a manufacturing business depends upon its employees and the need for learning and innovation in current practices.

Promotes 100% participation

The Lean TPM approach promotes improvement at the point of activity. It secures the engagement of the entire workforce in change and innovation in terms of thinking about how work is conducted, identifying waste and workers as the source of new ideas, new ways of working and sustainable improvements. Whilst some businesses display posters stating that 'Employees are our Number One asset', Lean TPM companies create a compelling vision to truly engage their employees and reinforce the involvement of the workforce by getting them involved rather than clinging to mantras.

Supports organisational learning

Combining efficient design with a focus for organisational learning provides access to and new more effective ways of working especially given the power of TPM. TPM has the proven power to break through the learning barriers that have prevented a meaningful optimisation of the manufacturing process and upskilling of operator teams to engage in greater diagnostic improvements related to the assets they control.

Incorporates a comprehensive loss measurement system

One further aspect of the Lean TPM approach to which we have alluded to during this chapter but not yet explored is a very powerful Lean TPM measurement system. This measurement system goes beyond the traditional measures of manufacturing. It provides visibility of previously hidden management losses in such areas as planning processes, new product development and technology losses and unmet customer needs. The measurement system tells each manager how far the business has progressed and whether improvement activities are generating an increased competitive capability for the firm. To date there are many examples of 'kamikaze improvements' that have glittered for a while and released absolutely no benefits to the firm or its customers. The Lean TPM approach is not so

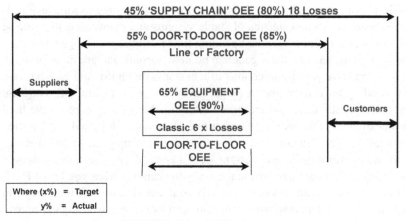

FIGURE 1.5 Lean TPM measures.

forgiving. It is not a blunt measure nor is it one that can easily be 'manipulated' as so many previous measures of 'world class' performance have been.

At the basic level there is the analysis and trend information that relates to a single asset or cell. This has been referred to as the overall equipment effectiveness (OEE) measure (Nakajima, 1986) or the 'floor-to-floor' level of analysis (see Chapter 3 for calculation). This aspect of the Lean TPM measure shows how well a machine/cell is managed. The disadvantages of the measurement include the ignoring of the chain of machines that supply (or take products) from the asset and form the internal production chain. Hence, the 'door-to-door' measure includes these linkages. Finally, at the highest level of control and the level at which 'manufacturing' can be exploited, as a means of competitive advantage, is the 'supply chain OEE'. These are interrelated measures, giving different levels of management analysis and trend information, covering the entire production system. These measures can be understood by all managers and can be used to target improvement/optimisation efforts and rid the firm of waste. We will return to the use of these measures and their role in the optimisation process in later chapters (Figure 1.5).

The measurement system therefore allows for proper navigation of the firm starting with the optimisation of individual assets, the optimisation of the chain of assets in the factory that form the production sequence and finally the overall performance of the firm and its selected supply chain design.

1.7 CHAPTER SUMMARY: THE FOUNDATION FOR A BETTER IMPROVEMENT MODEL

Modern competition calls for a new model to the generation of added value and profit. This is a 'moving feast' so the model must be robust and capable of continuous improvement if the efforts of everyone in a company is to keep pace with customer expectation. The consequences of 'falling behind' in this race are catastrophic for the organisation and every employee.

Improving the quality and delivery processes of the organisation and leading a business to a new agenda of stable systems and innovations to optimise the business is the new agenda. Spending time in endless meetings that generate no actions, have no data, generate no new learning and are more political than progressive are all indications of a business that must change – because it is heading for catastrophe. A progressive management team will recognise the power of the team and enjoy working together as they take on the hard new challenges of the market. Competitive advantage is designed and it results from the design of the organisation – that is quite simply the combination of staff and innovations for products, processes and staff. By itself, technology is not a form of competitive advantage – anyone can buy what you have! Products can be copied and patents avoided and if you deskill staff then they can be copied too! But an empowered and engaged workforce, working to improve processes and innovate with products, is the recipe for success.

Innovation is fostered and it earns its return when ideas are commercialised – all 'world class' performers know this and exploit it. This is the fight that keeps them going and the concern that remains in their minds even when they get to be the best in the market. If a business is not improving then it is in decline.

So the fight must include getting everyone to understand that they fight for the business and that 'old ways' must change to reach and sustain the highest level of OEE. That is the total response needed and the improvement of quality, improvement of flow/delivery and reduction of unnecessary costs pass through the business and its departments – these issues are not those of an individual. The customer does not care for departments and if department boundaries (that do not exist beyond the picture of the organisation chart) slow innovation then we must rethink our business because structure is inhibiting strategy (Figure 1.6)!

A Lean TPM business promotes people as innovators, to make suggestions and change systems that are failing. No one is offended if an employee asks a direct question or exposes problems with a graph! Participation is key and there

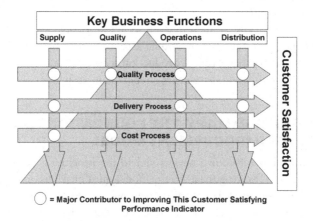

○ = Major Contributor to Improving This Customer Satisfying Performance Indicator

FIGURE 1.6 Cross-functional management.

is a two-way exchange of involvement and new skills. These skills are required to innovate and to add value and these skills must be nurtured.

What we have presented in this chapter is the beginning of a 'blueprint for change'. The mandate for change already exists for every manufacturer. The Lean TPM blueprint, whilst specific to each firm that uses it, has a power that is commercial and embodies the process optimisation and 'zero losses' needed to compete effectively and to engage change in a meaningful manner that is focused and understood by every employee. Lean TPM has a master plan that must be shared across the organisation and with suppliers/customers in order to optimise material flow and shorten the time between receiving an order and being paid for your efforts. This is the power of a high-performance manufacturing process where not a single moment of planned production is wasted.

REFERENCES

Brown, S. (1996). *Strategic manufacturing for competitive advantage*. London: Prentice-Hall.

Hill, T. (1985). *Manufacturing strategy*. Basingstoke MacMillan.

Liker, J., & Hoseus, M. (2008). *Toyota culture: The heart and soul of the Toyota Way*. London: McGraw-Hill.

Liker, J., & Convis, G. (2012). *The Toyota way to lean leadership*. London: McGraw-Hill.

Mann, D. (2010). *Creating a lean culture*. New York: Productivity Press.

Nakajima, S. (1986). *Introduction to TPM*. Portland, OR: Productivity Press.

Rich, N. (2002). Turning Japanese? PhD Thesis. Cardiff University.

Rother, M. (2010). *Toyota Kata: Managing people for improvement, adaptiveness and superior results*. New York: McGraw-Hill.

Schonbereger, R. (1986). *World class manufacturing: The lessons of simplicity applied*. New York: Free Press.

Schmenner, R. (2012). *Getting and staying productive: Applying swift even flow to practice*. Cambridge: Cambridge University Press.

Slack, N. (1991). *Manufacturing advantage*. London: Mercury Press.

Suzaki, K. (1987). *The new manufacturing challenge*. New York: Free Press.

Womack, J., & Jones, D. (1996). *Lean thinking*. New York: Simon and Schushter.

Womack, J., Jones, D., & Roos, D. (1990). *The machine that changed the world*. New York: Rawson Associates.

Chapter Two

The Lean TPM Master Plan

2.1 ACHIEVING THE RIGHT BALANCE

Lean TPM is one of the most powerful organisational transformation programmes of all. It combines and builds robustness into many other improvement programmes including the approach known as six sigma (an advanced approach to quality management) and builds robustness into the bufferless lean systems. As long ago as November 1997, at the TPM5 biannual conference of European TPM practitioners, Professor Daniel T Jones addressed the conference delegates on the topic of lean thinking and TPM. His observations were that although Just-in-Time is an accepted concept, most industries still scheduled work through departments in batches, worked to forecast and sold from stock, had long lead times, high buffers and poor quality detection. These are key target improvement areas for 'lean production' and the lean enterprise business model. To the casual observer, the lean approach has a different emphasis to the classic TPM focus on equipment reliability. There is some overlap, but together these cover 12 different target areas. So why would a recognised leader of 'lean thinking' be talking at a TPM conference?

The common thread is that both TPM and lean manufacturing highlight areas of historically accepted or hidden wastes (Womack & Jones, 1996). Despite their different origins, progress with either depends upon sensitising the organisation to recognise wasteful behaviours and practices. In effect, these improvement programmes create a heightened sensitivity to these 'wastes' so that each employee can detect the slightest of deviations in the production process and identify these as abnormal and take appropriate actions to restore production. Such an approach makes employees quite intolerant to other organisations that still maintain old business models and have not yet engaged in this form of waste elimination. In the case of TPM, the root cause of this waste is a short-term management perspective that tolerates and accepts poor reliability. The root cause of lean wastes is optimising parts of, rather than, the total value stream. TPM companies have always channelled improved effectiveness to increase customer value, but lean thinking helps to sharpen the definition of value. Lean thinking has always sought reliable processes, but TPM provides the route map to zero breakdowns and continuous improvement in equipment optimisation. Lean efforts without TPM are unreliable and TPM without the lean logic improves efficiency but may not translate this into customer value and improved cash flow.

The penultimate slide in Dan Jones's presentation showed the potential gains from lean as reducing:

- throughput time and defects by 90%;
- inventories by 75%;
- space and unit costs by 50%.

Overall, this potential to double output and productivity with the same head count at very little capital cost could equally be presented as the potential of TPM. Both Lean and TPM have evolved in parallel from their early concepts and are converging towards a common goal. But who cares? As long as there are benefits, all ideas are welcome. To understand what these are, it is worth taking a brief journey through the origins of both lean thinking and TPM.

2.2 THE ORIGINS OF LEAN THINKING

Inventor and entrepreneur Sakichi Toyoda and his son Kiichiro Toyoda, the family that founded the Toyota Motor Corporation, began to produce weaving looms and then cars in the 1930s. The approach taken by the family was to engage in a variant of flow production (Ohno, 1988a) that later matured to become known as the Toyota Production System (TPS) and has more recently become known as 'lean production' (Monden, 1983). At the heart of the manufacturing system was an attention to using simple machinery that automatically stopped, and assembly lines that could be stopped by operators, when a defect was detected (a system known as *Jidoka*). In this manner, no defective products would be passed forward to internal customer operations (Shingo, 1981; Schmenner, 2012).

In the West, traditional large batch sizes (and the responsibility for quality inspection being the role of a specialist department only) meant that defects could move within a factory and be hidden in buffers only to generate interruptions downstream (as defects created earlier were detected later and costly rework undertaken). Other factors conspired against the development of the mass production system at Toyota not least the lack of natural resources and large amounts of capital to fund investments in large-scale and dedicated technology – certainly in the post World War II (WWII) era. To counteract the lack of resources, Toyota engaged a production system that did not rely upon forecasts for each department but used a pull system (Ohno, 1988b) and enlisted the thinking talents of all employees to seek ways of improving. The TPS is also known as the 'Thinking Person's System' for this very reason.

Under the lean system parts are made on demand or as soon as the customer orders one unless it is not possible to achieve this form of fast flow (called Just-in-Time production). When it is not possible to satisfy a customer instantly then a pull system is used. A pull system operates with a small and set amount of materials which is there for customers (internal or external) to remove stock and satisfy their demand. The action of a customer removing the stock creates a pull to the earlier production stage to replenish the consumed stock. As such, production is strictly controlled and standardised inventory buffers deliberately disconnect operations. In this way, the movement of production materials from a supplier to a customer operation created a replenishment order – no computer was needed because this was a physical activity when stock was taken and the replenishment was triggered and empty boxes with control cards attached to them (called kanban) were sent to the production unit for refilling.

The basic pull system was later supplemented with a deliberate approach to level the workload of each production area (called *Heijunka*). It was not until after WWII that Taiichi Ohno (Toyota's Chief Engineer) compiled these practices to form the lean TPS that exists today (Womack & Jones, 1996) and a production system that operates at a very high level of efficiency and effectiveness (Womack, Jones, & Roos, 1990).

Ohno-san was a man with a vision and the architect of the TPS. He was a man with a strict intolerance to production wastes and a passion to optimise the flow of work – the cash-to-cash cycle. His intolerance has led to many great stories about Ohno-san and lots of stock would allegedly send him into a complete temper rage! His intent was to introduce a production system of high variety production in small volumes. Such an approach was therefore completely at odds with the Western passion for large batch sizes, dedicated and expensive technology and forecasting all operations under an era of mass production. He commenced the production system for engine manufacturing (where he had served his time as a Toyota employee) before extending it to vehicle assembly and later to include all Toyota suppliers (during the 1970s). In effect, Toyota now had a total 'pull system' network of materials supply that allowed instant availability of materials and a system that worked to replenish (pull) what had been consumed rather than pushing huge batches through operations to meet estimated forecasts.

Figure 2.1 shows how a pull system works with a limited amount of finished products held to allow immediate customer satisfaction. Upon consumption, the inventory level drops and this causes an internal order (kanban card) to be returned to the drilling operation. The drilling line then takes products, from a controlled kanban stock, and drills the hole in the product. As the inventory between the drilling line and the diecasting operations falls, this places an order on the diecasters and the return of a kanban card. The diecasters replenish the orders and so on through the manufacturing process.

The traditional approach is to use a **push** system and starts with a forecast by the company or order provided by the customer for products at a certain date in time (Rich, 1999). Let us say the 27th of the month. If it takes 2 days to drill the holes in the product, then that means the products must arrive on the 25th of the month, and if it takes 3 days to diecast, then the products must be launched on the 22nd of the month. So the physical manufacturing process starts with the release of materials in the diecast section on the 22nd of the month and the pushing of the products to the drilling station to meet the deadline of the 25th and then the push through drilling to the customer. That differentiates a push from a pull system at the most basic level. Obviously, any slippage in the push system, such as a machine breakdown, will stop the flow and immediate satisfaction of the customer, whereas under the pull system, the standard kanban inventory provides customer service and affords some protection from internal process disruptions.

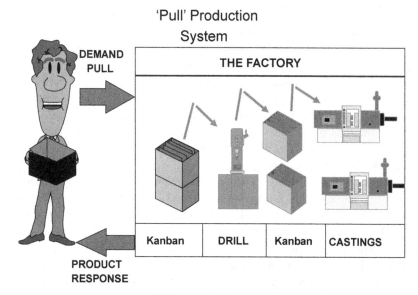

FIGURE 2.1 Pull system.

To support the pull production approach, Ohno and Toyota had also engaged in widespread quality management processes using employee involvement and had focused these activities on eliminating 'waste' from factory activities (Ohno, 1988b). The great quality guru Dr Edwards Deming echoed these principles in his work on industrial quality management (Deming, 1986) and when combined with the quality practices of Dr Juran created a powerful mechanism for change (and without doubt these two gurus helped shape the system known as policy deployment). In this manner, Toyota had, rather than concentrating on making machines process quicker (and engage in batch production with high buffers and poor quality), changed the emphasis towards the redesign of the production system for faultless production. Such a design allowed the elimination of all the poor features of mass production that generated excessive production costs and slowed the flow of materials in the factory. These forms of factory waste were identified as seven key aspects of production management – 'the seven wastes' (Table 2.1). By concentrating upon the elimination of waste, the amount of value-adding time improves as materials do not sit in buffers or contain defects. As a result, flow performance improves and the cycle between paying for materials, conversion and sale of the materials is compressed to improve the operational and financial performance of the firm.

The seven wastes have also been summarised into the word **TIMWOOD** which stands for wastes and costs associated with Transportation, Inventory, Motion, Waiting, Overproduction, Overprocessing and Defects.

After an extensive study funded by the Western automotive industry (IMVP Study), the performance advantage of Japanese automotive manufacturers was recorded in the publication *The Machine That Changed the World* (Womack et al., 1990). It was this publication that first used the term 'lean production'

TABLE 2.1 The Toyota Seven Wastes

The Waste of **Overproduction** where vast amounts of products are made in batches and simply 'dumped' into finished goods or work-in-process and result when there is a mismatch between customer demand for products and the ability of the production system to make to that demand. This is one of the greatest problems with mass production and the reliance upon large batch sizes.

The Waste of **Unnecessary Inventory** where the results of overproduction and other 'unimproved' constraints means that inventory is simply held awaiting an order in the belief that future orders will come.

Inappropriate Processing is another waste that results from a mismatch between the processes needed to make a product and the processes that are in place. In this manner, many firms use very sophisticated machinery to manufacture simple products that would be best produced using 'simpler' and less-expensive technology. Typically, in the West, large sophisticated machines with high processing speeds tend to be 'pumped full' of production in order to ensure a 'payback' for the asset and keeping such machines occupied with work inflates batch sizes and generates inventory (two forms of waste).

Unnecessary Transportation. A further form of waste concerns the movement of materials around a factory from the receiving back to the shipping bay. This activity can consume many hours and involve many kilometres of transport (with each activity offering the potential for product damage).

Unnecessary Delay concerns the simple 'dwelling' time as products are ready to be converted but sit waiting. For much of factory time, materials will be 'idly hanging around' in an uncontrolled manner.

Unnecessary Defects is the production of materials (that consumes value-adding time) but have to be reworked or scrapped. In this way, valuable capacity is lost forever – you cannot reclaim it even by working overtime. So imagine the problem of large batch sizes, long travel distances and, hidden within these batches, defective products!

Unnecessary Motion occurs when the production process is poorly designed and operators engage in stressful activities to handle materials. This is an unusual waste (ergonomics) but as claims for repetitive strain injury rise, many firms are facing large settlement fees from employee claims and solicitor bills.

to describe a new form of manufacturing developed by Toyota and adopted by most Japanese assemblers. The estimated advantage, resulting from the factory benchmarking process, was concluded to be a Japanese advantage of 2:1 in productivity terms and nearer 100:1 in quality of vehicle build (see Figure 2.2). The basis of this 'manufacturing mastery' was found to be a tightly integrated and synchronised manufacturing and supply system that, due to the lack of stock buffers in the total system of material flow, was termed 'lean production'. To put it another way, the Japanese producers could make products in half the time of the West and enjoyed the benefits of near-perfect materials entering the vehicle build process (measured in terms of parts per million defects rather than percentages).

In a later study of automotive component manufacturing plants in the UK and Japan (Andersen Consulting, 1993), 5 factories, of 18 in the survey, were deemed to be 'world class' and managed to generate high levels of quality and productivity simultaneously (see Figure 2.3). These factories were Japanese but not all Japanese managed to achieve the 'world class' status. Interestingly, when a line of 'best fit' is drawn between the non-world-class companies (Japanese and British) it shows an old engineering adage – that you can have productivity

Indicator	Japanese in Japan	All Europe
Performance		
productivity (hours/car)	16.8	36.2
Quality (defects/100 cars)	60	97
Layout		
space (sq. ft/car/year)	5.7	7.8
Inventory (sample 8 parts)	0.2	2.0
Size of repair area		
(% assembly hall)	4.1	14.4
Workforce		
% in teams	69.3%	0.6%
Suggestions/employee	61.6	0.4
Absenteeism	5%	12.1%
Training of new production		
workers (hours)	380.3	173.3

FIGURE 2.2 The benchmarking findings (Womack et al., 1990).

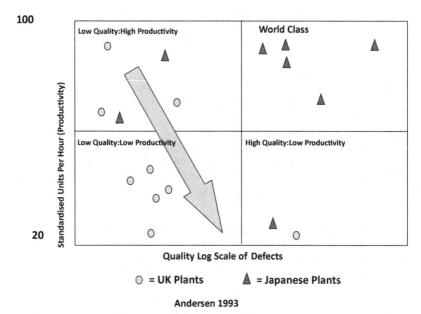

FIGURE 2.3 Andersen benchmarking findings.

but you will generate poor quality or that you can have slower production but good quality. This ability of 'world class' companies to break this trade-off using lean production principles firmly established the lean approach to manufacturing as an alternative design to that of traditional mass production.

In 1994, the study was repeated (same products) but included a wider sample of countries involved (9 countries and 81 manufacturing sites). The survey results, whilst less dramatic than its predecessor, revealed an 'underperformance' by Western firms and a dominance of Japanese firms (and emulators of lean production). This dominance encompassed all measures of 'world class' performance (quality, delivery, costs and people metrics). This dominance of producers all exhibited the same type of manufacturing system, they were all 'lean producers' and these studies firmly founded the lean model as an alternative to the 'ails and weaknesses' of traditional mass production designs (Andersen Consulting, 1994). These surveys also confirmed the competitive power of lean systems especially for manufacturers in very competitive and harsh markets and the ability of lean producers to maintain the highest levels of customer service with low levels of inventory buffering. Since this time the automotive sector has widely adopted lean systems and many more sectors beyond the context of car production have made significant gains from the application of the approach.

Lean thinking: Beyond auto production

The power of the lean production system design did not escape the attention of companies in other industrial sectors and many Western manufacturers, in industries far removed from car production, began to adopt aspects of the lean production model and to enjoy performance improvement. These businesses included aluminium-converting companies, aerospace businesses and general manufacturers. The importation of the lean production model, using its logic rather than a simple copying of techniques, was termed *Lean Thinking* and was popularised by Womack and Jones (1996) in a book with that title. The book examined over 50 cases of emulation and demonstrated 'before and after' comparisons of performance improvement. The authors also set out five basic pillars of lean thinking that, when implemented in order, generated the foundation for high-performance lean manufacturing.

From this time onwards, the terms 'lean production' and the 'lean enterprise' entered popular management terminology and have continued to redefine the model of the post-mass organisation. Lean production and the emulation of lean systems is now practised in a wide variety of industrial sectors, by large and small companies, and is rapidly being transferred to the supply bases of companies 'going lean' (Table 2.2).

Many businesses have now engaged with lean experiments and some now have mature lean systems which deal with the delivery of orders in a timely manner. However, a good industrial engineering approach to the business is not enough to ensure high performance. A lean maintenance system is needed. A business that does not integrate its maintenance skills with all other functions (purchasing, operations, logistics and others) will be severely limited and if an asset (machine, crane or vehicle) was to fail then, with minimal stocks in the system, the system will come to a halt very quickly. Also, without TPM it is impossible to meet the demands of a Just-in-Time system of delivering up to eight times a day to the line side of the customer. Great quality and delivery performance means control – control means the design of systems that ensure the reliability, robustness and

TABLE 2.2 The Principles of Lean Thinking

1. **Understand Value** in terms of 'WHAT' the customer wants to buy and what provides customer satisfaction/ customer service. This stage includes understanding the wastes in the current production system that stop or delay the process of information and material movements to provide ultimate levels of customer value. A general 'rule of thumb' used widely suggests that materials in a production system spend less than 5% of the time having value added (converted ready for sale rather than delayed, transported, etc. which simply adds costs and no real value for customers).

2. **Identify the Value Stream** and the internal activities undertaken within the firm that converts a customer order into a fulfilled order and the activities associated with generating new products for customers. Once you understand how you manufacture and design products you can improve the process and from here you can begin to work with the wider value stream (suppliers and customers) to eliminate all the wastes between companies involved with satisfying customers.

3. **Make Products Flow** is the third pillar of lean thinking and involves keeping materials and information moving so that materials 'flow' to customers without delay or interruption. Stocking materials for very long periods of time reduces stock turns and this inflates costs and ties up huge amounts of capital in materials that are not being sold for a profit.

4. **Pull Production at the Rate of Consumption** is used when it is not possible to completely flow products to customers (due to the number of customers, short lead times, the needs of your technology and batch sizes or other constraints). Under this rule, where it is not possible to flow production, a buffer must be deliberately designed to allow customer orders to be fulfilled from a carefully managed stock point. In this way, it is possible to maintain customer service by later production and finishing processing, 'pulling' out the work they need to complete orders from this buffer point. For advanced forms of lean production, it is possible to have many small buffer points that are used to directly link internal customer and supplier production operations and allow customer orders (removed from finished goods stocks) to completely pull work through the factory. This is known as the kanban system at Toyota and allows instant availability of products and short lead times simultaneously.

5. **Seek Perfection** in every aspect of the business and its relations with customers and suppliers is the final pillar and rule of lean thinking. Here, the authors stress the use of problem-solving teams of operators, managers and intercompany teams to squeeze out the last remaining elements of waste and non-value-added activity.

resilience of the production system. Lean without effective maintenance systems is an irresponsible management design that is unlikely to support a lean organisation over the medium term. Also, it is only possible to destock a manufacturing business and offer the production of small batches (ideally one order at a time) if processes are designed for reliability and there are enough maintenance procedures around them to detect the slightest sign of variation or loss of control.

2.3 THE ORIGINS OF TPM

The planned approach to preventative maintenance was introduced to Japan from the United States in the 1950s. Seiichi Nakajima of the Japan Institute of Plant Maintenance (JIPM) is credited with pioneering the development of the approach through the stages of preventative (time-based) maintenance, productive (predictive/condition-based) maintenance and then into total productive maintenance (Nakajima, 1988).

The JIPM went on to identify the following five critical success factors for delivering benefits from TPM:
- maximise equipment effectiveness;
- develop a system of productive maintenance for the life of the equipment;
- involve all departments that plan, design, use or maintain equipment in implementing TPM;
- actively involve all employees from top management to shop floor workers;
- promote TPM through motivation management: autonomous small group activities.

This focus on total involvement and motivational management brings a new perspective towards equipment management at all levels of the business. If a customer has only 2 h of stock then it figures that no supplier could have a breakdown of longer than 2 h. An unreliable process will increase stocks and will consume cash to invest in stock. The truth uncovered through TPM is that if equipment fails to deliver its 100% potential, it is due to some physical phenomena which can be identified, brought under control, reduced and possibly even eliminated.

The simple TPM goal of 'improving equipment effectiveness by engaging all those who impact on it in small group activities' is supported by a powerful business- led cross-functional improvement process which is easily missed by the casual observer.

Firstly, the Key Performance Indicator (KPI) effectiveness is a measure of how well a system works compared to expectations. Classic TPM identified six *effectiveness losses* which account for the gap between current effectiveness levels and 100% effectiveness. These are:
- breakdowns due to equipment failures;
- set-up and unnecessary adjustments;
- idling and minor stops;
- running at reduced speed;
- start-up losses;
- rework and scrap.

Each of these is a different type of problem with a different set of tactics to reduce and eliminate that loss. From the development of these tactics, the JIPM also identified that the *main reasons for poor levels of effectiveness* are:
- equipment condition is poor;
- human error/lack of motivation;
- lack of understanding of how to achieve optimum conditions.

This leads to one of the most powerful insights gained from the application of TPM principles. That the quickest way to improve performance is to recognise that there are only five basic solutions to any manufacturing problem. These are set out in Table 2.3 as a hierarchy of solutions starting at the base with low cost or no cost solution of setting standards or guidelines and progressing to the most expensive at the top of capital investment. This hierarchy sets out the sequence of steps to delivery problem prevention rather than problem solving.

TABLE 2.3 TPM Problem-Solving Hierarchy

	Solution	Format	Added Value
5	Apply major capex project	Improve customer value	Deliver step-out products and services
4	Apply low-cost automation	Reduce non-value-added activities	Improve flow and flexibility
3	Refine process control	Optimise process controls	Improve mean time between intervention and reduce defects
2	Improve a best practice	How to carry out an activity	Improve work methods and competencies
1	Define/refine a standard or policy guideline	When this happens, do that	Improve visibility of standard

Although at first glance, this may seem oversimplistic, it is the mastery of these five steps that will lead to stable and then optimum operational effectiveness. Levels 1 and 2 are low-cost or no-cost solutions which resolve 9 out of 10 of the root causes of sporadic failures. This is why the foundation steps of TPM are to improve equipment condition (solution 1: what standard should be set for equipment condition) and to standardise practices for correct operation and asset care.

The elegance of this approach is that to achieve steps 1 and 2 it is necessary to define what should be done and to ensure that people are competent to do it. Once the first two steps are completed, the tactics developed for each of the six losses can be applied in turn to systematically release the full potential of the asset.

This process is the core ingredient of the TPM master plan which sets out how:

- routine tasks can be systematically transferred from specialists or management tasks to front-line teams;
- specialist and management roles evolve to support the delivery of strategic goals.

Those who understand the full potential of TPM refer to it as Total Productive Manufacturing and at least one European organisation (Heineken) refers to it as Total Productive Management. In Lean TPM, Total Productive Manufacturing combines the power of lean manufacturing and Total Productive Manufacturing as equal partners on the journey to industry leading performance. The earliest example of TPM used in Europe, to deliver world class performance, was in Belgium where Volvo won the PM prize for their work in the paint shop in Ghent. TPM was adopted quickly by a number of automotive companies as they took massive action to try and catch up with Japanese levels of quality and productivity. In the United Kingdom, the potential of TPM really became recognised in the early 1990s – a full 20 years after Toyota and its extended supply base had begun their TPM journey. Where it was implemented correctly, it produced benefits for individuals as well as company profitability. Like most lean practices some observers thought the practices were restricted and culturally specific to Japan – but this was simply untrue. The challenges posed by equipment the world over mean managers and teams must find countermeasures such as TPM – the manner in which the system will be implemented will differ by culture.

A trade union leader presented another landmark statement. Writing at the time, Sir Ken Jackson, then General Secretary of the AEEU commented, 'We recognised the value of TPM some years ago. We saw then that TPM could enable manufacturing and services to become the best in the world. Unlike TQM, which was conceptually sound, but patchy in outcome, TPM offers a new and invigorating approach. Involving everyone from shop floor to boardroom, TPM is a team-based and freshly focussed tool for success'. Such an endorsement by one of Britain's most progressive trade unions is important and underlines the 'growth potential' associated with the development of operations personnel and increasing their skill sets to take on higher levels of self-management in the factory.

The evolution of TPM

Classic TPM evolved in line with the changing needs of industry which led to an extension of the scope of improvement beyond the production shop floor and across supply chains. This occurred at a time when lean thinking was emerging to challenge traditional mass production concepts and link supply chains more closely through flow. The combination of these emerging ideas plus the success of TPM at improving equipment resilience led to the extension of classic TPM scope. Company-wide TPM includes administration, safety, environmental, quality maintenance, product development, customer service and logistics. It also led to the rationalisation of improvement tools and policy deployment under the focused improvement pillar. This extension of TPM provided a means of aligning improvement across the organisation to get everyone involved in a focused and meaningful manner. As part of this enhancement, the range of hidden losses was extended to 16 to include a wider range of hidden loss targets covering equipment, labour, management and energy/resource losses.

2.4 LEAN TPM

Lean TPM is a logic, a methodology and a set of integrated techniques that release the customer-facing commercial gains of the lean approach through the robust processes of TPM. The Lean TPM is more than just a simple addition – Lean TPM is an evolution and natural result of combining the knowledge and learning organisation based on high-performance industrial engineering with the knowledge and learning from scientific engineering processes. The synergy and 'multiplier' effect is far more powerful than any other programme. Lean without TPM is a bufferless system that is exposed to process failures and these can be highly expensive if customers impose penalty charges for non-delivery of their orders. TPM without Lean will often lead to large warehouses full of stock that consumes lots of cash and may not add any value for customers or dismantle the 'large batch thinking' of the past. Starting either programme leads to change and learning and this brings both improvement programmes closer and closer together. The combined logic is extremely powerful and everyone benefits – customers, managers, office staff, specialist maintainers/engineers, other specialists and operator teams. A Lean TPM business is a vibrant one, it is improving on a daily basis, it is checking on a daily basis and it is hyper sensitive to the merest sign that the process is varying beyond acceptable standard – finally the organisation is learning every day no matter how many years into the programme.

The key to releasing the power of the Lean TPM business model is leadership – leaders set direction and have the power to overturn old habits and behaviours. Leadership changes culture – and culture is 'the way we do things around here'.

The Lean TPM Top-Down management role

Only management have the opportunity to set and enforce standards/policy to address equipment losses and bad behaviours. It is they who define priorities, allocate resources and counteract unacceptable behaviour. The top-down role is to define priorities, set standards and provide recognition to reinforce proactive behaviours. Under Lean TPM, the management team adopt the following five principles as the foundation of their management outlook, which are as follows:

- Continuously improve overall equipment effectiveness by attacking the six losses through small cross-functional group activity.
- Establish standard practices for correct operation incorporating routine asset care by operators.
- Establish processes to sustain and improve equipment performance over the lifetime of the asset.
- Increase the skills and motivation of all those who impact on equipment effectiveness.
- Apply early management techniques for new products, services and capital projects to deliver flawless operation from day 1 and ongoing low operational life cycle costs.

Although simple, these principles, when applied together, trigger a profound series of transformations across the organisation. The first principle is supported by the delivery of improvement through front-line cross-functional teams. Their priority is to formalise and stabilise shop floor processes. In addition to the expected gains in quality, cost, productivity, delivery, safety and morale, practices are simplified and delegated to front-line teams. As these teams develop improvement capability and the organisation is able to progress from solving problems to optimising value stream performance, the senior management leadership role changes.

The Lean TPM master plan sets out a series of linked learning landscapes through which senior management can develop organisational capability and deliver industry leading levels of performance (Table 2.4).

TABLE 2.4 Lean TPM Master Plan Milestones

Milestone	Improvement Learning Landscape
1	Establish basic flow processes, remove excess inventory. Set standards for equipment, processes and behaviours. Formalise current practices against them.
2	Refine flow layout and inventory/planning parameters to reflect true demand profile. Simplify and refine core activities and remove sources of accelerated deterioration to deliver the 'zero breakdown' goal.
3	Extend time between interventions (reel-to-reel running) and improve end-to-end supply chain collaboration. Improve project management of products, services and capital projects to achieve flawless operation from day 1 production.
4	Condition way of working to sustain process optimisation activities. Focus on step-out products and services to set the customer agenda. Reset the forward master plan based to ratchet up customer value and sustain step-out performance.

Although this is set out as a linear process, progress through each improvement landscape is a voyage of discovery. Senior managers that navigate each transitional learning landscape well are the ones that master the skills of:

- Change leadership: Make it clear where the organisation is going, what change is needed and how fast, what are the change milestones, who will be involved, how to manage progress.
- Innovation: Stimulate new thinking by setting cross-functional step-out goals and accountabilities.
- Alliance development: Build on internal collaboration by extending working relationships beyond company boundaries to create new value, reduce transaction costs and accelerate growth.

The senior management leadership process is characterised by:

1. Setting accountabilities for direct reports to deliver improvement through front-line cross-functional teams.
2. Allocating front-line teams a budget of around 5% of their time to deliver improvement within their improvement zone.
3. Establishing underpinning processes to
   a. Monitor team progress with a formal feedback every 3 months.
   b. Support the development of teams, team leaders and managers to meet their improvement accountabilities.
4. Holding managers and teams accountable and reinforcing progress with team-based recognition processes.

The quarterly feedback provides the heart beat for reporting demonstrated results, sharing best practice/knowledge and refocusing on team goals. Align the feedback timing with the business planning cycle as part of the policy deployment process. The importance of the team-based improvement process cannot be over emphasised. This is the vehicle to establish and reinforce proactive relationships across functions. Although only 5% of time is used, the impact on engagement and motivation is far reaching.

Behavioural benchmarks are summarised in the Team Review and Coaching (TRAC) framework graphic in Figure 2.4. TRAC benchmarks integrate the principles and values of Lean and TPM into a single collaborative team capability development framework.

Level 1A benchmarks provide an indication of the development of basic improvement team capabilities including understanding of process steps, identification of top improvement targets and development of plans to deliver. Capability is measured by the demonstrated delivery of those plans and realisation of performance goals. As the TRAC audit/coaching process measures progress through the capability levels, it also identifies team development needs. Once level 1A is achieved, levels 1B to 4B are used to progressively guide the development of high-performing team capabilities. The role of front-line manager is to coach their teams so the use of TRAC also provided a structured leadership development path for team managers at all levels.

Finally, the collective results of all teams provide feedback on progress against the Lean TPM master plan. Levels 1A and 1B relate to milestone 1, 2A and 2B to milestone 2, etc.

The Lean TPM Bottom-Up role

As mentioned above Lean TPM is a cross-functional team-based process improvement activity that may only take up a small percentage of attendance time but it is an essential element of engagement and ownership. This is the vehicle by which front-line personnel can identify a purpose to believe in and be provided with the support and reinforcement to deliver their full potential.

The TRAC stepwise benchmarks guide the development of cross-functional collaboration and high-performance operations teams. Progress through each step heightens everyone's sensitivity to abnormalities in the workplace and breaks down traditional barriers. Those who attempt to shortcut this process by cherry picking readymade practices from other organisations will achieve little in the way of engagement or bottom-line benefits.

The TRAC framework incorporates the steps of autonomous maintenance interlinked with the corresponding planned maintenance, OEE improvement and education and training activities. It also incorporates the relevant lean manufacturing activities to support value stream flow, flexibility and focused improvement.

This stepwise process has the effect of raising the capability of production teams, maintenance teams and supervisors and of releasing specialist resources to focus on the next development stage. That is process optimisation, a key part of the 'proactive maintainers' role once breakdowns are brought under control. This point is a switch where staff experts can release their time to engage in improvement activities because their days are not beset with catastrophic failures. It is therefore possible to spend 40% of the time engaged in standard daily management activities and 60% on improvement projects.

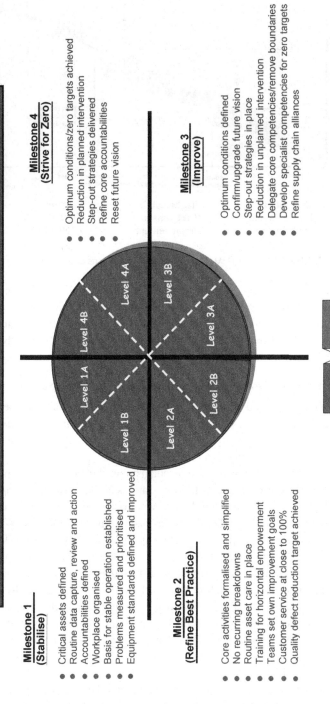

Bottom-Up TRAC Framework

Milestone 4
(Strive for Zero)

- Optimum conditions/zero targets achieved
- Reduction in planned intervention
- Step-out strategies delivered
- Refine core accountabilities
- Reset future vision

Milestone 3
(Improve)

- Optimum conditions defined
- Confirm/upgrade future vision
- Step-out strategies in place
- Reduction in unplanned intervention
- Delegate core competencies/remove boundaries
- Develop specialist competencies for zero targets
- Refine supply chain alliances

Milestone 1
(Stabilise)

- Critical assets defined
- Routine data capture, review and action
- Accountabilities defined
- Workplace organised
- Basis for stable operation established
- Problems measured and prioritised
- Equipment standards defined and improved

Milestone 2
(Refine Best Practice)

- Core activities formalised and simplified
- No recurring breakdowns
- Routine asset care in place
- Training for horizontal empowerment
- Teams set own improvement goals
- Customer service at close to 100%
- Quality defect reduction target achieved

FIGURE 2.4 Team review and coaching framework.

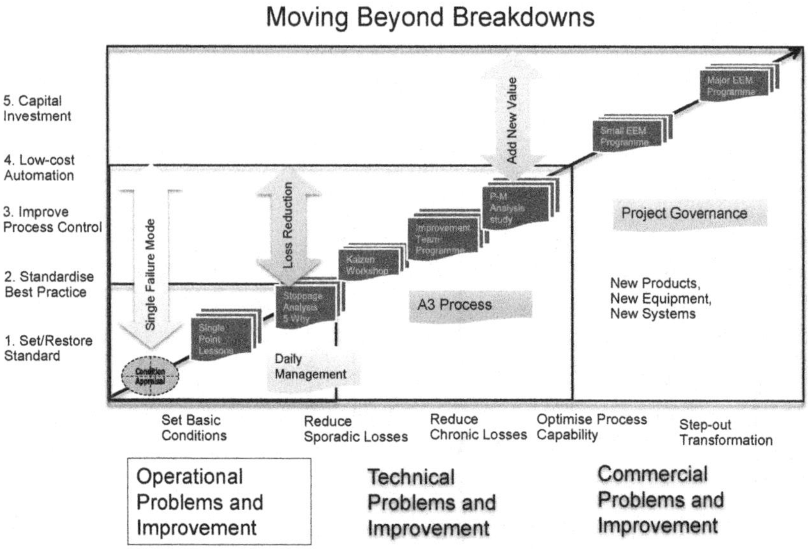

Moving Beyond Breakdowns

FIGURE 2.5 The TPM journey to zero breakdowns and beyond.

The Lean TPM journey includes the use of a range of improvement tools which firstly establish basic conditions to stabilise processes and then progress to optimise process and value stream capabilities. As can be seen from Figure 2.5, this follows the hierarchy of five basic solutions introduced in Table 2.3 above.

The Lean TPM master plan: a blueprint for change

As set out above, the top-down role is to lead the organisation through a series of learning landscapes. This is supported by a stepwise bottom-up development process that builds capability, improves ownership/morale and increases production capacity (time that can be sold to customers).

The Lean TPM master plan is the mechanism used to co-ordinate top-down leadership and bottom-up delivery. The foundations of this master plan approach were developed within TPM to guide the evolution of operator and maintainer engagement process. In Lean TPM, this has been refined to guide the 3–5-year business transformation process and produce operations capable of world-leading levels of performance (Akao, 1989).

This approach is based on the road maps used by successful and award-winning companies and research conducted by the Lean Enterprise Research Centre. These are road maps developed in response to the changing leadership and organisational learning challenges presented as each layer of waste is 'peeled back'. By setting these out as learning landscapes, our aim is to help management to recognise and prepare for those challenges, minimising the risk of inertia and loss of direction.

The master plan is therefore the combination and sequence of activities that form the logic and plan for the Lean TPM implementation. Unsurprisingly this also defines the scope and priorities applied through the policy deployment process discussed in the next chapter.

The overview of each master plan milestone sets out how Lean and TPM principles and techniques are applied to guide progress across each learning landscape. This is set out in this way so that readers experienced in either Lean or TPM can see how the combined approach adds value.

It is a feature of those who have developed their improvement spurs through lean to see TPM as simply operator maintenance. It is a feature of those who have developed their improvement ideas through TPM to see lean as simply set-up reduction and supply chain planning. This is a mistake of ego. A limiting behaviour which unfortunately we are hard wired to repeat. Unless you have tried to apply these tools, your frame of reference is too narrow. If you have progressed this far through the book, you are searching for answers. Many are not as open to new ideas as you which is why the step prior to embarking on the master plan is to carry out a pilot programme.

The goal of the pilot process is not to prove that Lean and TPM work, but to understand how to adapt your organisation to benefit from it. This is best achieved through a practical but limited project over at least 3 months. This will provide the insight into the process gaps and barriers to change. Just like the back to the floor TV series, you need to first understand the shop floor reality. Only then can you be sure you have the right frame of reference to move forward. 'Time spent in reconnaissance is seldom wasted' (Sun Tsu 544 BC).

Pilot/launch; learning to see

Aim: To achieve senior management consensus concerning the future vision, assess the current benchmark/potential and provide the hands-on model and experience necessary to populate the first two master plan milestones and develop a realistic and achievable plan to mobilise the journey to world class levels of performance.

- Management awareness of competitive conditions and competitor analysis to focus the appropriate response and direction of change needed by the firm.
- Refine measurement process, carry out scoping study, gap analysis and accountabilities.
- Macro mapping to identify base case and improvement potential.
- Confirm strategic vision/goals and define in detail the operation needed to deliver those goals.
- Align short-, medium- and long-term goals under a single change agenda/ master plan.
- Establish change infrastructure including recognition systems and motivational levers.
- Develop a 12-month plan in detail including logistics of releasing people for training and to support the programme.
- Apply Lean TPM to pilot value streams/centres of excellence.
- Company-wide engagement with 5S/Can Do.

MILESTONE 1: ROLL-OUT CASCADE
(INTEGRATING THE INTERNAL VALUE STREAM)

Aim: To establish the company-wide 'best-practice recipe' for low inventory, high flow and stable operation.

Bottom-up: first-line management led
TPM tasks:
- Communicate commitment and gain 'buy-in'
- Establish teams and build plans
- Awareness raising and TPM education
- Effectiveness measurement
- Equipment condition
- Understand and reduce scattering of dust and dirt (contamination)
- Asset care/maintenance process
- Formalise operations start-up, steady state, closedown routines

Lean tasks:
- Understanding the value stream and system of production
- Process flow value mapping, identification and elimination of waste
- Flow alignment/cell creation
- Stabilise inventory levels
- Non-value-added removal

Top-down: management led

Formalise best-practice standards to guide progress towards master plan goals such as zero breakdowns, class 'A' planning system that ensures all bills of materials, routings, inventory and forecasts are correct to achieve 100% customer service.

Exit criteria (how to identify when the capability to progress further has been reached)
- All employees involved with high levels of first-line management (FLM) ownership.
- Performance gaps are assessed to focus improvements. Improvement activities undertaken to deliver stable operations.
- Factory contains showcases of good practice.
- Technical maintenance, industrial engineering and quality management records are reviewed, streamlined and compiled to form an effective management reporting system.
- The costs of poor performance are assessed and tracked to demonstrate the relationship between the ownership of improvement and waste (cost) reductions.

MILESTONE 2: REFINE BEST PRACTICE
(MAKE PRODUCT FLOW)

Aim: To 'lock in' the control systems for low inventory, high-flow operation delivering zero breakdowns and self-managed teamwork.

Bottom-up: FLM led
TPM focus – floor-to-floor losses

- Simplify and consolidate maintenance tasks to reduce technical judgement. Introduce single point lessons (SPLs)[1] and review process documentation for efficiency.
- Achieve real empowerment/shift beliefs by engaging in further rounds of structured problem-solving including detailed analyses of cause-and-effect relationships together with the inclusion of mistake-proofing devices to prevent problem recurrence.
- Achieve cross-functional shared ownership of assets between operating shifts. Also the establishment of common practices such that innovation is shared, through standard process documentation, as the platform for new improvements.
- Clarity concerning operations 'trouble map' so that improvements become focused on removing greater levels of waste.

Lean focus – door-to-door losses

- Compress the internal value stream by colocation of machinery to form cells where possible.
- Focus on quality of product to identify and react to the production of defective materials. Defect detection prompts problem-solving activities by teams. Focus on quality is intended to raise productivity and lower costs whilst also reducing batch sizes and inventory needed to support the operation.
- Common measures of performance displayed in factory areas including measures of morale, safety, quality, delivery and cost reductions achieved by the teams.
- Integration of support departments especially those engaged in tooling manufacture and departments affecting the new product introduction process.
- New policies affecting the allocation of maintenance spares, location and control of these items. Also strategies for sub-assets are determined effectively (i.e. repair versus replace cycles).

Top-down: management led

- Improve underpinning training and increase diagnostic tools available to improvement personnel.
- Align activities and share results internally through presentations and copying of practices between areas with similar issues.
- Transfer roles, new skills sets and change job descriptions. Establish competencies records to show employees who have received training (internal and external) and what level of achievement they have reached (i.e. trained, capable of training or expert).
- Integrate shop floor team outputs into planning process and capital projects.
- Prepare vision alignment to make use of increased capability/lead time.

---

[1] A single point lesson (SPL) is a single piece of A4 documentation that contains all the information needed to learn or conduct an operations/maintenance task safely and efficiently. These documents often include flowcharts to represent procedures, machine drawings and/or digital photographs to show the procedure simply.

Exit criteria (how to identify when the capability to progress further has been reached)

- No recurring problems and greatly extended time between maintenance interventions.
- Stable operating conditions and predictable usage of consumable and spares items.
- Routine self-managed maintenance and operations combined within a structure of SPLs that are reviewed by teams to ensure they represent good safe working practice and are efficient. An SPL is a single A4 paper document that explains all the major features of a process or task so that anyone can follow and understand it.
- Recognition structure for contribution of teams, suggestions for improvements, training and performance (safety and efficiency).
- Factory environment containing high levels of visual management (colour coding) and maximum use of communication boards containing key information about company and area performance.
- Impromptu and formalised problem-solving groups.
- Documentation of the production system and integration of the logic and techniques of the production system integrated with employee induction on 'on the job' training.

MILESTONE 3: BUILD CAPABILITY (EXTEND FLOW SYSTEMS)

Aim: To identify the recipe to release the full potential of the current operation and build the foundations to match and exceed future customer expectations. This includes raising standards to deliver outstanding performance and the flexibility to deliver world-leading levels of performance.

Bottom-up: FLM led

TPM focus

- Identify zero target priorities to optimise customer value.
- Define and control parameters to optimise manufacturing progress.
- Reduce labour intervention with machinery and need to adjust machinery to maintain quality performance. Significant progress towards 'no touch' production releasing operations staff to engage in project work and problem-solving with a greater technical content.
- Raise shop floor team competence/capability to deliver self-management.

Lean focus

- Lead time reduction activities and time compression from the value chain of operations.
- Statistical process control engaged at all critical assets to detect, predict and control production. Self-recording by local area teams.
- Quick changeover between products and strategy to achieve 'single minute' or 'one touch' set-ups for maximum product variety to be manufactured at each stage of the production process.

- Production teams and support staff engage in the analysis of competitor products.
- Self-certification of suppliers and integration of strategic suppliers with the operations system. Including levelling of demand for materials to avoid traditional problems associated with poor forecasting accuracy of production schedules (lumpy demand where the production facility would have little to do than be beyond its capacity in a very short timeframe. This is also known as 'boom and bust' production).

Top-down: management led

- Focus on competitors and how to 'step change' the internal production system to levels needed to compete effectively in the future. Extension of thinking to cover the next 5 years and not just the current day.
- Setting the competitive agenda and predicting and promoting future customer needs.
- Promoting innovation and establishment of key cross-functional business improvement initiatives including a focus on the supply chain and integration of supplier businesses needed to enhance material flow. Integration and development of engineering services and spares providers.

Exit criteria (how to identify when the capability to progress further has been reached)

- Have identified critical optimisation targets (commercial, operations and technical) and making progress towards them.
- Maintained zero breakdowns.
- Established a clear product/service 'innovation stream' strategy in place capable of achieving customer leading performance.
- Focus shifted from internal improvement to include external partners.
- Technical focus shifts to external scanning for asset innovations rather than internal correction. Engineering staff to ensure that they are included in any future asset specifications at the capital expenditure and procurement stages. Feedback by improvement teams to this knowledge base is routine. This includes intelligence about improvements needed for the next generation of assets.

MILESTONE 4: STRIVE FOR ZERO (PERFECTION)

Aim: To change the competitive landscape and set the future customer agenda for products and services.

Bottom-up: FLM led
TPM focus

- Optimisation of asset performance.
- Flawless integration of new technology (including commissioning).
- Increased focus on knowledge management/cross-project learning.
- Flexible production without labour constraints.
- Zero quality defects and losses.
- No boundaries between functions.
- Introduction of reliability-centred maintenance and condition-based monitoring procedures.

Lean focus – supply chain loss

- Establishment of fully integrated supply chain, speed of supply and pull system of materials with customers and suppliers. Integration of the pull system with customers creates a form of dependency and affords some protection from competition.
- Re-engineered supplier evaluation processes and integration of suppliers with business strategy sharing, key change programmes and widespread development/best practice sharing. Rationalisation of suppliers to eliminate excessive numbers of alternative suppliers or to create 'systems' suppliers providing ranges of products or configuring entire product subsystems rather than just offering components.
- Development of concurrent product development with suppliers. Sharing of resources between companies including 'resident engineers' at supplier factories.
- Common logistics systems and use of common logistics providers (use of Just-in-Time deliveries and milk rounds for product collection).
- 'Right sizing' of tooling and assets to meet life cycle needs of product. Simplification of machinery (avoidance of procuring unproven technology). Maintenance and life cycle cost of ownership routines fully developed and integrated with product costing systems.
- The weight loss of converting raw materials to finished products examined to find methods of reducing the amount of conversion necessary. Integration of suppliers with near-net-weight programmes.
- Full integration and standardised process for the introduction of new products in short cycle times. Mass customisation of products to allow a logical variety of products offered to the customer but with minimum disruption to the production process.
- Environmental effectiveness of the organisation and its technology prioritised as a key competitive capability.
- Business engages in extensive networking with other businesses in nonrelated fields to seek out new innovations.

Top-down: management led

- Consolidation of optimisation gains and focus on steps to control competitive agenda.
- Internal improvement resources and trained personnel directed to assist suppliers and customers with improvement activities.
- Increased deployment of business costs to shop floor teams. Increased integration of teams with the execution of business policy deployment. Increasing responsibility of factory teams to propose key changes and present these to management as annual themes or key projects to deliver market success.

Exit criteria (how to identify when the capability to progress further has been reached)

- Delivering industry leading standards of delivery of customer value.
- Established strategy to disrupt current competitive landscape and control the rate of change for other organisations in the same sectors and market segments.
- World class standards of product innovation and customisation strategies.

2.5 WHAT DOES LEAN TPM OFFER?

Lean thinking methods and approach improves the design efficiency of the transformation process providing the potential to deliver greater customer value with less effort. This includes frameworks to identify 'customer winning' patterns of operation and opportunities to secure competitive advantage. TPM tools improve the effectiveness of the transformation process (i.e. dealing with the reasons why things do not go to plan). This includes frameworks to release capacity and increase control and repeatability. The implementation process is designed to change attitudes, to change relationships by increasing the dependency of employees on each other, to develop capability and increase cross-company collaboration for improvement. Below in Table 2.5 are examples of how Lean TPM provides a holistic approach to continuous improvement driven by progressively removing inhibitors/barriers to change within the business and across the complete supply chain.

Using Lean TPM to challenge current thinking and clarify business drivers is an important part of the implementation route map (which is why everyone in a business should be involved with the programme) and this provides a practical way of making business drivers and their links with continuous improvement visible. All business functions can then target continuous improvement of business drivers, making it part of the day-to-day routine.

TABLE 2.5 The Benefits of Lean TPM

Measure	Impact of TPM	Impact of Lean Thinking
Productivity	Reduce need for intervention Reduce breakdowns	Reduce non-value-adding activities, increase added value per labour hour
Quality	Potential to reduce tolerance Control of technology Reduce start-up loss	Highlight quality defects early
Cost	Reduce material, spares	Lower inventories
Delivery	Zero breakdowns predictability	Shorter lead times, faster conversion processes
Safety	Less unplanned events Less intervention Controlled wear Minimise human error	Less movement, less clutter Abnormal conditions become visible easily
Morale	Better understanding of technology More time to manage	Less clutter Closer to the customer Higher appreciation of what constitutes customer value
Environment	Closer control of equipment Less unplanned events/human error	No 'overproduction' Systems geared to needs not theoretical batching rules

The result is an increase in the range of options available to a business when responding to economic change and a key management resource supporting the drive to sustain competitive advantage. Or to put it another way, Lean TPM has a direct impact on better quality, better dependability, shorter lead times, greater flexibility and lower costs – essentially everything a modern market demands. Lean TPM is the only controlled approach to reducing stocks and allowing a quality-assured process to be 'speeded up'. Speeding up a process that lacks a robust maintenance system simply serves to move defects around a manufacturing system by workers who lack the skills to detect the abnormalities in the process that indicate the process is becoming unreliable. Allowing defects to move along a manufacturing process (with the potential that they escape to be detected by the customer) serves to generate noise and interruption to flow as well as reducing the image and relationship of the business with its customer base.

The policy deployment process will therefore tend to adopt Lean TPM as a systematic methodology to stabilise the production process in the short term and then to engage in improvements that take the business closer to its future competitive state. The policy deployment process will also provide the platform to understand how Lean TPM can generate value now and in the future. The ability to treat the business as a single entity and to promote the interests of the business (and everyone's long-run pension generated by adding value profitably) places the loyalty of manager on the level of the business rather than the myopic and dysfunctional traditional focus on the interests of the business department. In this manner, much more meaningful discussion can happen in terms of eliminating the problems of existing machinery when buying the next generation of assets (early management), by commissioning assets quickly (often known as vertical start-up or loss-free start-up) and ensuring assets are optimised over their working lives (lowest life cycle costs). Management discussions – during the catch ball process – are also likely to involve better and more effective product designs to all design for manufacture, design for assembly and better environmental manufacturing through effective and lean product designs. The discussion will also highlight one fundamentally important activity and company rule – every competent employee must have the authority and responsibility to stop the production line as soon as they detect a defect or before when the process is about to lose control.

2.6 TACKLING THE HIDDEN WASTE TREASURE MAP

The most important leverage to support successful waste reduction (the objective of early-stage Lean TPM) is clarity of accountabilities. Accountabilities drive priorities, so if accountabilities do not change progress will be short-lived. The concepts of equipment floor-to-floor (F2F) – effectively the individual item of equipment's performance measure, door-to-door (D2D) – the measure from goods inwards

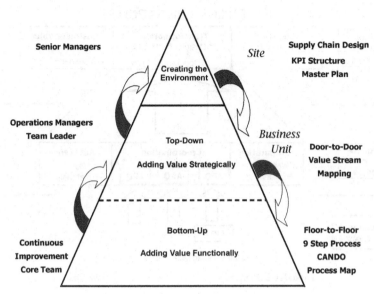

FIGURE 2.6 The organisational structure of Lean TPM.

to finished goods, and the supply chain measure of intercompany performance link together to form a complete value stream measurement system. As shown in Figure 2.6 these accountabilities align with short, medium and long term planning horizons. As such they clarify and reinforce management roles and provide a framework to align priorities across all three levels. This then becomes a core part of the performance management framework linking top-down priorities with bottom-up shift team delivery. Accountabilities are set relative to the decision horizon, which each level of the organisation controls (short, medium and long term). Success at each level is, therefore, linked to the effectiveness of activities in the other two.

The design of the hidden loss treasure map is based on the following definitions of Effectiveness hidden losses. By that, we mean things that go wrong but should not prevent us from achieving the full potential of the process.

1. *Availability losses* relate to issues which prevent the task from starting.
2. *Performance losses* relate to issues which reduce the output of the conversion process once under way.
3. *Quality losses* relate to issues that reduce the quality of the output.

Analysing the effectiveness of processes at each management level helps to identify the nature of the hidden loss. Is the management system or process broken or just damaged? Is there potential to speed it up and can we increase precision? Each of these areas of hidden loss can be further categorised in terms of:

- *Unplanned losses* usually due to the lack of a standard or guidelines to recognise or prevent occurrence.
- *Systematic or planned losses* usually resolved by improving how the work is organised.

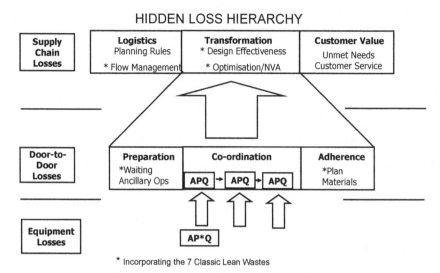

FIGURE 2.7 Measuring Lean TPM losses and wastes.

The resulting categories provide a way of assessing areas of strength and weakness at each management level in a way that also helps to what issues exist. Applying these definitions to each of the three levels of effectiveness creates a treasure map containing 18 hidden loss categories or wastes as shown in Figure 2.7. (see above). Like a map, this provides a guide to the territory rather than a detailed replica of the actual landscape. The treasure map framework combines the seven lean wastes with the six TPM losses. It also identifies other hidden loss areas relating to logistics, co-ordination, creating customer value and design processes.

The treasure map helps to identify organisational effectiveness losses and highlights cause and effect mechanisms across organisational levels. This insight can then be incorporated into the policy deployment process to align improvement activities at all three levels. With clear accountabilities, when the team succeed their value to the organisation can be recognised and savings established. This reinforces the right values, beliefs, profit improvement focus and the central concern of all manufacturers – to raise customer service. These losses are discussed in more detail in Chapter 4.

2.7 CHAPTER SUMMARY

Both Lean and TPM have evolved in parallel from their early concepts and are converging towards a common goal (Rich, 2002). Both are company-wide approaches and not narrow sets of techniques. Some have suggested that 'lean tool heads' do exist and they sell techniques and this is true – 'tool heads' miss the links that create company-wide and co-orchestrated 'systems' change (Seddon, 2006). Lean and TPM have both achieved significant results by delivering practical solutions to different business issues. 'Lean thinking' has tools to design efficient supply chains (Womack & Jones, 1996). TPM has tools to improve supply chain effectiveness (Willmott & McCarthy, 2000). The combination of these approaches improves both operational efficiency and organisational effectiveness. Without this focus it would be all too easy to make improvements but not to convert this efficiency into cash by lowering inventory buffers in a controlled manner. In this respect, a staged approach to implementation must be adopted to avoid 'kamikaze improvements' that generate more heat than improvement. To achieve improvements, it is imperative that the approaches are not defined narrowly and that a cross-functional management infrastructure is created to ensure the benefits of the change programme can be exploited properly.

The implementation route map provides the change process to align accountabilities and progressively ratchet up operational capability. Such a route map helps to co-ordinate the application of lean thinking and TPM tools and techniques to secure continuous improvement in business performance in terms of quality, cost and delivery and the methods needed to remain competitive in today's intense markets.

Much of the preceding chapters have been dominated, implicitly and explicitly, with the importance of being visual when managing and also establishing roles and responsibilities for planning and daily management. Lean leadership typically uses visual management when aligning daily management and the breakthrough activities of the policy deployment process – this is often represented by a physical space in the factory known as an Oobeya room or 'big room' – sometimes called a 'war room' in the West. Toyota have made use of such rooms for many decades and used them as planning and review centres and an area within which all managers can understand the current status of projects and the business by actually having desks in the room concerned or being brought there for very regular reviews. The Oobeya room is therefore critical to maintaining the collaborative focus of managers and allows different specialists to engage with each other to achieve the Lean TPM master plan that has been selected during the policy deployment process. A dedicated area for senior and middle management is a further demonstration of the importance of the change programme and the need for a strict observance of the plan–do–check and act cycle of senior management.

REFERENCES

Akao, Y. (1989). *Hoshin kanri*. Portland, OR: Productivity Press.

Andersen Consulting. (1993). *The lean enterprise benchmarking report*. London: Andersen Consulting.

Andersen Consulting. (1994). *The second lean enterprise world-wide benchmarking report*. London: Andersen Consulting.

Deming, W. E. (1986). *Out of the crisis*. Cambridge, MA: Center for Advanced Engineering MIT.

Monden, Y. (1983). *Toyota production system*. Atlanta: Institute of Industrial Engineers.

Nakajima, S. (1988). *Introduction to TPM*. Portland, OR: Productivity Press.

Ohno, T. (1988a). *JIT for today and tomorrow*. Portland, OR: Productivity Press.

Ohno, T. (1988b). *Toyota production system: Beyond large-scale production*. Portland, OR: Productivity Press.

Rich, N. (1999). *TPM: The lean approach*. Liverpool: Liverpool University Press.

Rich, N. (2002). *Turning Japanese?* PhD thesis. Cardiff University.

Schmenner, R. (2012). *Swift, even flow*. Cambridge: Cambridge University Press.

Seddon, J. (2006). *Freedom from command and control*. Buckingham: Vanguard Publishing.

Shingo, S. (1981). *A study of the toyota production system*. Tokyo: Japan Management Association.

Willmott, P., & McCarthy, D. (2000). *TPM: A route to world class performance*. London: Butterworth Heinemann.

Womack, J., Jones, D., & Roos, D. (1990). *The machine that changed the world*. New York: Rawson Associates.

Womack, J., & Jones, D. (1996). *Lean thinking*. New York: Simon & Schushter.

Chapter Three

Policy Deployment: Aligning People, Processes and Products Profitably

3.1 TRANSLATING DIRECTION INTO FORWARD TRACTION

Chapter 1 should have convinced you there are many reasons to change and build a business that is fit for the future. Chapter 2 sets out Lean TPM and why a plan for the future must be based on good leadership, engagement of staff and a structured approach to deliver the full potential of all employees and full commercial benefits. By now you will see that the first two Lean TPM milestones take a business from its current state to the path to stability and then how milestones 3 and 4 set out the path to optimised operations. The length of time this journey will take depends upon the organisation concerned so it is important to see the journey as a commitment.

A 5-year planning horizon for the master plan is long enough to change the direction of most businesses but is close enough to be able to predict most obstacles and trends in markets with a good degree of certainty. Senior managers can use this to think through what will happen to markets, new competitors and technology over the long term to set the future direction of the business. So the commercial plans naturally combine with the master plan unlocking the commercial benefits for the business to compete.

Having set out the common vision for change (what we want to achieve and why) and how to develop the capability for change (the master plan), our attention now needs to be directed towards how to coordinate the actionable next steps to make it happen.

To deliver the promise of the master plan, it is not appropriate to just manage delays, errors and failures in the current business. This is like looking a few metres beyond the front of your vehicle whilst driving forward. It is not enough to see obstacles and opportunities coming your way, and it is a sign that

Lean TPM.

the business is not planning and redesigning its processes to deliver value. It is important to look forward when using Lean TPM methods – a planning system that looks forward should be capable of foreseeing issues (positive and negative) in the market and how tuning the Lean TPM system can be adapted to provide the competitive advantage in the future and in existing/new markets (Akao, 1989).

A process to align each master plan milestone with strategic business drivers and translate that into 'live' projects to prepare the business for its future is missing in most businesses. It is called Policy Deployment (PD) and it is a very special process.

Sounds great doesn't it? And it is. Despite airplanes full of Western executives taking the modern pilgrimage to Japan to tour Toyota and other great manufacturing businesses, many came back with stories of working practices they could see. These include the pull systems of stock that is removed by customer processes and eight deliveries a day of a small amount of product to the line side of the vehicle assembler. What people have seen on their tours is typically 'the visible' solutions to problems that were experienced or predicted but in Toyota's past. The pull system was the solution to offering a full availability of products to customers in a way that allows them to just come and take what they need. It was the solution to a system that would make only what is needed, and it is also the solution to a process that cannot and does not have the cycle time to supply orders on demand. Hence, the buffer to protect a slower process must remain in batch mode because it cannot supply on demand – this practice is known as a kanban in Japan.

Lots of executives came back with great stories of the visible solutions but they did not see the logic and principles of the system nor the hand that guides the system. The 'hand' is known by many names, not just Policy Deployment. In Japan, it is termed Hoshin Kanri which means 'directional control' or 'directional management' and is analogous to navigation of a pathway using a compass where the needle points 'due North' so a route can be mapped. In the West, it is probably better known as Strategy Deployment or Policy Deployment. In this book, we will refer to it as Policy Deployment.

3.2 A FOUNDATION OF TOTAL QUALITY MANAGEMENT

Policy Deployment approach and exactly who perfected the system are debatable but one thing is for sure it emerged during the late 1950s and early 1960s. It is probably not a coincidence that during this period in Japan, the ideas, concepts and methods of Total Quality Management (TQM) were being developed and embraced by, what would become, leading Japanese manufacturing businesses. At the spearhead of the Quality revolution was not a legion of Japanese figures – these would come later – but at the start (in the post–World War II Japanese reconstruction period) it was the American quality gurus of Dr Deming and Dr Juran (sponsored by the Japanese Union of Scientists and Engineers, JUSE) who were leading the revolution.

The message they spread was that quality must be company-wide, process driven, involve problem-solving and must be supported by a planning system which management must own and it was received warmly by businesses. TQM as a competitive weapon was absorbed by Japanese businesses with relish who took on-board all the key managerial learnings needed to improve processes and the organisation itself. Statistical process control (SPC) is great and a feature of Lean TPM but without Policy Deployment there is no future business.

At the time, Dr Deming was promoting his management thinking and Dr Juran (his colleague) was promoting a Quality Council that needed to focus on breakthrough quality and the vital few projects that would equip the business to deliver excellent customer service. The vital few were selected by management and based upon predicted/known weaknesses in business processes and the predicted issues that the business would face. The combination of both gurus provided the conditions for policy deployment to become an integrated management practice at the same time that the West remained measuring outputs and condemning to the bin all failures to meet the standard or returning them to production to be reworked (more added cost). Before we go any further with this chapter, it is wise to summarise the contributions of these two figures – their advice remains as pertinent today as it did in the 1960s.

Dr Deming

Deming was a tall and impressive figure of an American man, who arrived first onto Japanese soil and has contributed enormously to the modern quality agenda and the modern practice of quality assurance. He had been sent to assist the post-war recovery of Japan and to teach a nation how to manage process quality. But he did much more than promote his process of Plan Do Check and Act or his approach to SPC. He brought with him a management model that became so admired that the Japanese nation created a Prize in Deming's honour, the Deming Prize (one of the most revered and sought after awards in Japan). Deming and Juran were well known to each other – they were colleagues and both were trained by the eminent statistician Walter A. Shewhart of Bell Laboratories in the USA.

Deming's contributions to the Japanese understanding of quality management included the following:
- **The Plan–Do–Check (Study)–Act cycle of improvement** where all change facilitators (managers and team leaders) should follow the four-step process. The first step was to Plan the improvement process by checking the current state of performance and what is needed to improve this condition. The next stage is to organise and conduct the improvement (Do) using teams and a systematic approach to result in the one or many activities needed to make the difference. The Study/Check stage is associated with the monitoring of the post-intervention and whether the team has made and sustained a difference. It was pointless to claim a victory if the improvement did not hold its gains. And finally there is Act where the teams must see how the improvement and its learning can be extended and replicated by other similar processes. These four stages form a process of learning and remains a

key element of lean practices today including the A3 improvement process and policy deployment itself. Policy deployment is the (PDCA) of planning. This approach was an adaptation of his former boss's approach known as the Shewhart cycle.

- **SPC** is probably (and sadly) the one thing that Deming is most remembered for. His approach allowed managers to observe processes and collect measures in a way that permitted the establishment of an average value (say the level of filling a bottle with water) and then through the natural variation of a process to show the range of values that would be collected. He used this range to mathematically establish control limits (measures that were used to trigger action if the value was found when an inspection was undertaken). Upon detection of unacceptable deviation, the process would be stopped and corrective action taken to restore what had gone wrong.

- **The Deming Seven Deadly Diseases** is a set of poor management practices that prevent businesses from achieving high performance and ultimately achieving TQM. The seven diseases include the following:

  - A lack of constancy of purpose by management and therefore the appearance that the manager of a business is jumping between reactive things to do to counteract problems with the business. Deming was keen to promote a constancy of purpose based on a focus on the customer and the quality, delivery and cost of what a business does.

  - An overemphasis on short-term profits which places management attention on the short term to the neglect of longer-term planning for the continued success of the business. Also a short-term financial myopia generates dysfunctions as staff compromise what they do to continually make short-term profit improvements (such as buying from the cheapest poor-quality suppliers to theoretically lower costs).

  - The use of an individual-based evaluation of performance, merit rating or annual review of performance. The focus on management by objectives (MBO) does not typically represent anything more than a boss telling a subordinate to implement a change (which will be reviewed in a year's time). There is typically no focus on solving a problem, the innovative capabilities of the subordinate are not used, and the process is purely one of implementing solutions (based on what senior management want) for senior management who are often not best placed to determine what is best practice or needed.

  - The mobility of management is a problem in terms of constant leadership changes and changes in direction.

  - The overfocus on the visible figures of a company alone – this is the failure of management to look at other performance indicators including those related to the safety, morale and social responsibilities of the business.

  - The need to pay excessive medical costs through poor human factors, poor ergonomics and poor job design which results in repetitive strains,

'hospitalisation' and compensation claims from current and former employees who have been harmed by the poor design of work.
- The excessive costs of warranty payments and legal payments paid to lawyers who earn their fees by following claims.

In addition to his seven diseases, Deming also noted that other management practices, approaches and beliefs could cause other 'lesser' problems and these included a neglect of long-range business planning, a management focus on emulating others rather than solving their own business problems, believing that automation will lead to improvement, using the excuse that the business is somehow different and therefore unable to learn or could be 'excused' as unable to achieve high performance, blaming workers for errors rather than believing that the system design is responsible for 85% of all problems, an overreliance on inspection to capture quality failures without using this information to feed back into better and more robust designs and not keeping up with management practices and knowledge of how best to manage.

- **The Deming 14 Points** – probably the most powerful of all of Deming's contributions was his approach to management and what he later described as his 14 points (Deming, 1986, pp. 23–24). The points reflect his general approach to systems thinking and how best to get an organisation (as a system) to act and behave in a way that maximised its outputs. It is certain that these principles were written and published in 1986, and reflect his approach to management which began and developed from his years in Japan. These points underpinned his activities with Japanese manufacturers and the lessons he taught to executives at what would become some of the world's most renowned producers and manufacturing brands. These 14 points and the previous methods are critical to understanding the process we now know as Policy Deployment.

His 14 points, paraphrased, for high-performance TQM are:

1. Create constancy of purpose, by managers, toward the improvements of the product and its processes so that the organisation becomes more competitive, and to do this all employees must become customer focused.
2. Adopt the new philosophy and adopt a new leadership style to break the traditional behaviours that were appropriate for the previous era of competition but not fitting for modern business. Leaders must lead during a period of significant change.
3. Businesses must cease their dependence on inspecting products and services and invest in building quality into products.
4. Stop doing business on price alone – price is not the total cost of a product or service. So cheap prices from suppliers of poor-quality products will need sorting out of defects before use or in production – these issues must be sorted and corrected and they involve additional costs to the customer business! Companies should take a total cost approach to doing business and limit sources of supply to only proven and trusted suppliers of the lowest total cost products.

5. Businesses must develop processes and a workforce who engage with continuous improvement of quality, productivity and to reduce costs.
6. Companies should introduce a system of on-the-job training that is conducted in a scientific method for the instruction and standardised work for employees.
7. Institute leadership at various levels of the business in a way that fosters employee innovation.
8. Drive out fear and accept that employees should acknowledge when things have gone wrong, and managers must design systems to learn from these failures and prevent them from happening again. Hiding problems from the fear that acknowledging failures will lead to discipline of the individual must be avoided.
9. Organisations are made up of processes rather than departments and barriers between departments must be broken down so that all employees regard themselves as part of a team. Only by acting as a team will the product and work be optimised.
10. Eliminate slogans and avoid slogans that urge the workforce to achieve goals such as 'zero defects' or improve productivity; these antagonise workers because they need major organisational changes to achieve and they are unlikely to be achieved by any individual or operational team.
11. Create pride in the work for all staff by focusing on the quality of work.
12. Remove MBO because it acts as a barrier to stop management and engineering innovations.
13. Create and continuously improve the education and self-improvement systems of the business so that skills are improved to add greater value.
14. Focus everybody in the company to work on the transformation of the business and make it clear how everyone fits with the business strategy.

Whilst Deming contributed a number of other major advances to the world and practice of TQM, these are his main contributions – all of which point to a systems approach where quality planning and improvement are linked with a management structure that promotes collaboration, constant learning and continuous improvement.

The 14 points are very interesting and remain as important today as they were in 1986 when Deming first published them. His impact on management thought has been profound and his ideas certainly underpin the policy deployment approach and so too does the guidance for quality offered by his colleague and friend Dr Joseph Juran.

Dr Juran

Dr Juran was also invited by the Japanese Scientists and Engineers to Japan to lecture business leaders on quality management in 1954. He was also trained by Walter Shewhart and was a believer in data-driven quality. His contributions include:

- **The supplier–process–customer concept** – Juran proposed that every department and individual in a business must understand that they are part of a chain within which they are a customer to someone else who supplies them and also a supplier to someone else. By stressing this dependency he believed you could generate meaningful improvements in business performance.

- **The Cost of Quality** – Juran was also a keen advocate of costing the quality process and he proposed that there were three levels of quality costs. The most expensive were when the quality system failed. Of less cost was when the system was operated using inspection and appraisal (checking for signs of failure then intervention to prevent the failure from happening) and finally the least expensive approach which was to design devices and processes that prevented failure from happening altogether.

- **The Pareto (80:20) Principle** – Juran drew inspiration from the economist Wilfredo Pareto. Pareto was the first to propose that most processes and activities are skewed in terms of volume and variety. As an example, 80% of all vehicles produced in the world are made by less than 20% of the total number of producers, or 20% of the total possible reasons for visiting an Accident and Emergency department will account for 80% of the visits to A&E. The rule is not hard and fast but it is a useful principle. For Juran it meant that management should focus on the vital few improvement projects – those that will bring the biggest benefits to the business – out of all of the possible projects that could be undertaken. These he considered to be 'breakthrough initiatives'.

- **Juran's 10 Steps** – Juran believed that TQM was a journey at two levels. The first he saw – in a similar way to Deming's PDCA/PDSA approach – as a journey of learning from symptom to cause and then from cause to remedy. At the second level he proposed a 10-step approach to organisational improvement. The steps are summarised as follows:
  - create organisation-wide awareness of the need and opportunities to improve business performance;
  - quantify the challenge and establish goals for improvement efforts;
  - structure the organisation and allocate effort to close the gap and reach the goal;
  - develop and deploy the necessary staff training and develop competence;
  - conduct projects and experiments to solve quality problems;
  - develop a regular reporting system;
  - recognise efforts and results of staff engagement;
  - promote the results and communicate progress throughout the organisation;
  - measure the progress so you know how you are doing; and
  - maintain the initiative and improvement momentum by establishing annual improvement challenges as a standard method for the company.

It can be no surprise that Juran's top–down approach was supported by his belief that each business should establish a quality council to control the company-wide improvement activities of a business. His 'trilogy' included the Planning of Quality, Quality Improvement and then Quality Control based on the management of 'breakthrough activities' and all co-ordinated by the council.

Juran, like Deming, provided many more contributions to quality management but these help position the Policy Deployment process as firmly founded upon the principles of quality and company-wide engagement in a process that is focused on the vital few projects that will equip the business to compete in the future.

3.3 THE POLICY DEPLOYMENT PROCESS

Before we explore the technical aspects of Policy Deployment, it is important that this highest level planning process (PDCA) is put in perspective. Policy Deployment is not a 'one-off' but instead it is systematised and fits with the financial planning cycles of the business (Juran's annual cycle). So, at Toyota, for example, the financial year runs from January to December and this represents the key management focus for annual cycles so it makes sense to align the planning process with heartbeat of the financial reporting/capital expenditure of the organisation.

So the financial year starts in January and this is *Step 1* – it is where all 'vital few' projects to improve the business have been 'signed off' and are ready to go. But the process starts well before this time. Officially the process starts in September of the previous year where the executives of the business establish the annual challenge (which they have derived from their research and studies which look at a 5-year and long-term planning horizon). So in September, the challenge is written down, evidenced and stated to the middle management team as a collective (by the senior management/executives). Often the challenges are expressed in output and customer terms so it will include a challenge to improve quality by 20% over 3 years of which the annual challenge is to achieve a 10% improvement, for delivery there will be targets for reduction in lead times and also in on-time in full and targets for cost reductions as well as staff engagement. This is particularly powerful – it is a business challenge to all managers in the business. This is not the dominance of marketing and sales telling everyone else what to do – this is what 'we' need to achieve to remain competitive and protect everyone's standard of living.

At this point, 3 months before the beginning of the new financial year, the 5-year view taken by senior management is expressed as an annual challenge. This could be presented to the assembled middle managers as an improvement of quality by 10%, a reduction in lead time, a cost reduction percentage and a goal to engage all of the workforce in the change process. The senior managers – after explaining the basis of their challenge – then retire and leave the middle managers to debate and negotiate the potential improvement projects that they could undertake. The purpose of this activity is to devise and agree the company-wide programme of change that will deliver the improvement sought for the business. The focus of the team during this stage – which is known as catch ball because it is similar to the child's game of throwing a ball between catchers – is to negotiate the 'hows' to satisfy the company need. The 'hows' will get very detailed so that everyone knows the implications of any proposed changes to 'the way we work around here'. Take, for example, the need to cut down batch sizes and make products more frequently in order to reduce the amount and therefore costs of finished goods and work in process stock. The team may debate the need for quicker changeovers to reduce batch sizes and allow the production planners to smooth demand and release stocks. The team will then debate what would be needed to achieve this quicker changeover and all the supporting activities (planning of production, human resources (HR) skills, tool preparation areas, common height tooling, pull signals from customers, etc.). Once agreed that the project is a viable option and an assessment of what it will bring in terms of 'payback,' the middle managers then acknowledge their agreement to the senior managers. They continue until they have exhausted the 'vital few projects'. Japanese businesses are very good at identifying five (give a few more or less) breakthrough programmes within which there will be subprojects. Western businesses tend to have a lot more individual-based programmes and therefore do not benefit from managing the process as a cross-functional system.

It is vitally important that the middle managers are used in this way – they are the most creative of all employees and they are the grade of employee who control systems within the organisation and tend to know what is the best practice. Senior managers add value by setting the direction of the business, and the shop floor teams add value by making products in a quality and high productivity manner and make a significant contribution to the improvement processes that control the daily management of making products and earning a profit. This approach to the division of labour is quite unusual and quite peculiar to the process of policy deployment.

The middle management team must agree and 'sign off' (sometimes quite literally signing the project charters) the breakthrough annual projects with the senior management team before the challenge and projects are communicated to the entire workforce, customers and suppliers. This must happen before 'Step 1' at the end of December – ready for the financial year. Often the teams of the factory are shown an affinity diagram which shows the drivers for change and the projects that have been selected – these pictures are then placed on walls throughout the factory to promote the annual challenge. It is also common to find the entire programme captured in a diagram known as an 'x' chart (Figure 3.1).

At Step 1, each team (with an assigned leader and membership and also a timeline for implementation) will present their team charters and then promote their project to the rest of the business (gaining vital feedback) before beginning the implementation process which also includes the 'bowling chart'. The 'bowling chart' which is a direct reference to that of the American 10-pin bowling sport is a record of performance which is based on the key measure of implementation used by each team. As in bowling each week represents a round of the game and the actual performance against the target performance for each measure is plotted (this can be trended against a time series chart if needed). Week by week the scores go up and to aid the reviewer each score is colour coded with green being the achievement of target, amber/orange being close to the

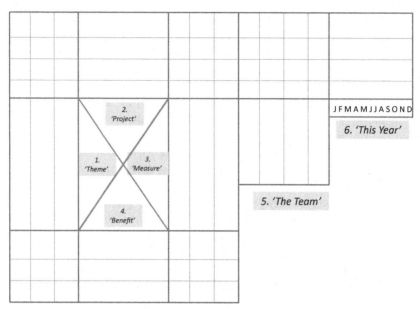

FIGURE 3.1 The X chart.

target and red being not close at all to target performance. As such these review measures, which are reviewed by the team and on a monthly basis by senior management, are easy to understand and draw attention to those measures and teams who are 'in the red' and may need help, additional resources or another team to deliver their part of the change programme. It should be noted that most but not all change programmes would have a numeric measure:

- employee engagement would be the percentage of all staff who have engaged with an improvement activity or undertaken training,
- timely deliveries would be an on time and in full measure,
- lead time would be from order placement to order fulfillment,
- quality would be the parts per million defects produced by suppliers, in process and at finished goods including the horrendous costs and reputational damage of returns from customers for quality reasons,
- flexibility of the production lines could be the maximum number of products (variety) that could be made in a day.

Some people disagree with the use of measures – but these are simply indicators of progress and the planned target achievement rate for the business to perform at a higher level of customer service and profitability. Measures must be carefully designed to produce the right behaviours so great care is taken to ensure system measures are compatible, measure progress and stimulate good learning.

Most programmes of change last only 9 months so they conclude towards the September date when the next challenge is about to be announced. This allows senior managers to reappraise the new current state of the business and set the new challenge and direction for the business.

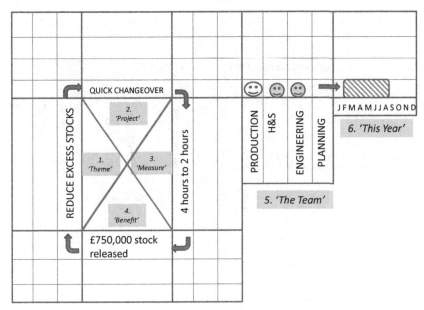

FIGURE 3.2 The X chart illustrated.

So that is the annual cycle for the business – financial systems and improvement systems running in parallel with the managers spending their time between daily management (working to lead their teams and looking for efficiency gains) and improvement/breakthrough management and the improvement for the effectiveness of the business system (Figure 3.2).

The 'X chart' is therefore completed for all the projects and programmes of change that have been agreed by the management team. The chart shows that the starting point is the theme to reduce stocks, the project is quick changeover, the target is to go from 4 to 2 h and this is expected to reduce stocks by £750,000. The chart also shows production, H&S and engineering will take part in the Quick Changeover project and that the Gantt chart shows a duration of February to August if you follow the project row across the chart to the right. The team can now add other projects for our theme and then add all other themes and projects (Figure 3.3).

3.4 THE CONTENT OF POLICY DEPLOYMENT

To add a bit more depth to the policy deployment system, it is important to review the system in terms of the content. There are many approaches to policy deployment but they each share a basic approach to setting a challenge and asking 'How?' as many times as needed to get a basic direction for the business all the way through to the exact changes needed to the production system. This filtering is conducted as the annual challenge is filtered through the many levels of management debate all the way through to the implementation team. I have often wondered if the story of the cleaner at the space agency NASA was part of

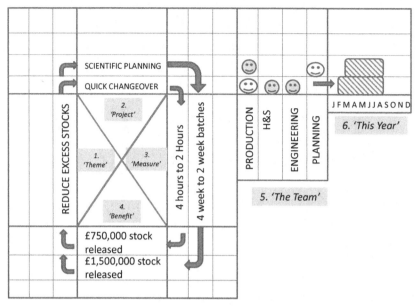

FIGURE 3.3 X chart expanded.

their deployment process – when asked what he or she was doing by a visitor, the cleaner replied I am keeping the facility clean and clear so we put a 'man on the moon!'. The logic is quite good – an orderly workplace is needed to ensure there is no noise and distraction to the value-added process of scientists working together to get an astronaut to the moon and that all the materials they need are close at hand. So if you have ever struggled with workplace discipline and have received the reaction that "'I cannot see the point in this!'" then perhaps you have not stressed the importance of getting everyone involved in the improvement process and it is your fault for not communicating the need properly. Incidentally, after thousands of company visits over the years, I have never seen a world class business with a shambolic working environment no matter how much or how little profit the business makes!

Back to policy deployment. The best way to explain the process is to follow a simple example. Let us consider the themes of change that are set out by senior management in 2015, which could be

- to increase the engagement of the workforce (with an associated target for 2018),
- to reduce the total costs of excessive stocks (with an associated target for 2018), and
- to improve the operational efficiency of the organisation (with an associated target for 2018).

Now we need to decompose these directional improvement areas, through study and observation of the business system, into more tangible annual improvements. Effectively we will take each theme and decompose it. If we take the reduction of excessive costs of stock, we may decide as a management team to investigate the reason why stock is carried in an amount that is far in excess of what is needed to satisfy customer demand. During this process, we may find that production batch sizes are not set with any real scientific understanding and are far too big, the reason why they are too big is through poor application of policies. We may also find that the equipment itself demands large batches because the changeover of the equipment is long and a lot of production time is lost through changeovers. So as a team we may decide to investigate the two key processes of the calculation of batch sizes and secondly an investment in a 'quick change-over of equipment' programme. The process of determining 'how' we achieve the desired direction of the business has resulted in two programmes for the challenge of stock reduction. Interestingly, the purpose of the catch ball approach is to pro-mote projects and it will be no surprise that the quicker changeover of machinery and better production planning will impact on operational efficiencies too.

It is now possible to establish – if the team has agreed to conduct a quick changeover programme – what actions would be necessary to achieve and improve but the team must now determine:

- What is the annual challenge for the measurement of changeover perfor-mance (so we can set up the bowling chart). Let's say the current perfor-mance level is 4 h of lost production, every changeover and the annual challenge is set to reduce the time taken to 2 h.
- Now we need to identify the staff who will form the programme team and lead the implementation of the change (production, engineering, health and safety, production planning, HR and quality assurance probably with the leader being the production manager).
- Now we need to know when this project will be undertaken. Sometimes the team will find another team's project must be completed first before their project can start. So let us say the project for quick changeover will start in February and end in August.
- Finally we can establish the financial savings that result from quicker changeovers at the 2-hour standard. So this would allow two more hours of production to be made per changeover, and it should mean that the produc-tion team can cycle through all the products they offer to customers every 2 weeks and not make a batch every month – this reduces stock and means that the finished goods levels and work in process can be lowered in a con-trolled way (possibly with the introduction of kanbans to make planning aligned with consumption).

This simplistic explanation should help you visualise how the process works. To aid reporting, the bowling chart is created so progress can be tracked in a very simple and visual way.

3.5 THE BOWLING CHART

The bowling chart is a simple chart which shows progress against plan. The chart can also be a time series chart. Please note the direction of change arrow and the happy face show that when this measure is trending in a downward direction the process is improving – this additional visual management of the measurement system allows the status of all improvement programmes to be seen at a glance (if they are presented on a wall in the management offices).

A business must decide how it intends to display the information concerning progress through a project/programme and also the achievement of the team. The main reason for plotting performance in this way is to ensure the teams are supported and timely action is taken by the senior management to support the teams. After all, the teams are working to perfect the future competitive advantage of the business and that is the responsibility of senior management. The two methods are shown below (Figures 3.4 and 3.5).

The important point is that progress can be seen and also that the team can show why a sudden improvement in performance has happened perhaps due to the redesign of the tooling used to make it mechanically easier to changeover.

The Lean TPM Master plan and policy deployment go hand in hand – policy deployment brings out the best of Lean TPM because it is a learning process and those departments that were traditionally disinterested in TPM can now understand it in terms that have meaning for them. So for sales staff the new approach means an opportunity to reduce stocks, improve the flexibility of the production lines to personalise products and offer customers the reliability they seek. The alignment of policy deployment with the master plan is critical if this learning is

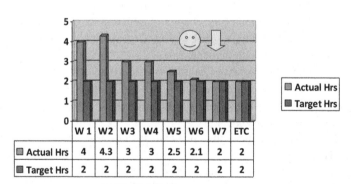

	W 1	W2	W3	W4	W5	W6	W7	ETC
Actual Hrs	4	4.3	3	3	2.5	2.1	2	2
Target Hrs	2	2	2	2	2	2	2	2

FIGURE 3.4 The bowling chart.

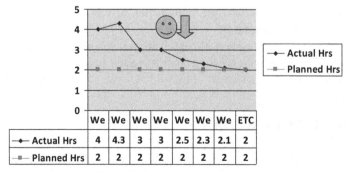

	We	We	We	We	We	We	We	ETC
Actual Hrs	4	4.3	3	3	2.5	2.3	2.1	2
Planned Hrs	2	2	2	2	2	2	2	2

FIGURE 3.5 The time series line.

to result in better engineered processes and to build a production system around the Lean TPM milestones. Policy deployment also lets departments such as the HR team to work on new contracts and new training programmes that are built ahead of time and built to meet the future needs of the business in skills that are specific to the business. However, there will be some companies that do not have the policy deployment system for their organisation – the policy deployment process can still be used but this time at the level of the Lean TPM team. At the very least it will allow the Lean TPM programme to be seen on one piece of paper and be controlled in a way that will help mangers of the production and support processes to understand the process and the timing of the milestones.

3.6 CHAPTER SUMMARY

If you believe that an organisation must evolve and adapt to compete then it follows that a business must become sensitive to what is happening in its market, internal process management and supply chain environments. By becoming more sensitive to environmental change and focusing on the direction of change so that a gap can be established between the current business performance and where it needs to be in the future is great and opens up so many more possibilities than traditional direction setting processes such as systems such as Management By Objectives (MBO).

The articulation and deployment of the future strategy by senior management, based on research and study, help to promote a common mental model of the future that all the business can rally around. It is very hard to argue against a well-crafted policy deployment document process that sets out the business case that will help every employee continue their employment. The policy deployment process also highlights the 'gap in predicted performance' to harness the talents of the middle management team to work together to find effective solutions. The outcome is a stream of linked challenges encouraging these managers to be innovative and solve the fundamental problems in the business that are stopping it from realising its vision. Policy deployment also clarifies where leaders add the most value. Typically this results in the development of Leader Standard Work to miminise the time a manager spends reviewing the day-to-day activities of the business releases valuable time for management of the policy deployment agenda. In fact, at Toyota in Japan, it is often taught that the daily work of directors should be routinsed to form only 20% of their time, for middle managers daily management should account for around 40% and for line team members daily management should account for 80% of the day job – the rest of the time at each level in the business is allocated to improvement and transformational activities. In this manner, daily work protects improvement work and makes it possible to use scare management resources as a group to maximum commercial impact. The potential of policy deployment is therefore a significant contributor to the competitiveness of the business, and it is the less visible hand of lean that guides the business to higher levels of performance, better revenues and total engagement of the workforce and supply base.

Policy deployment will naturally fit with the milestones of stabilisation and then optimisation as managers engage Lean TPM as the foundation for commercial success based on a highly reliable production system.

REFERENCES

Akao, Y. (1989). *Hoshin Kanri*. Portland, OR: Productivity Press.
Deming, W. E. (1986). *Out of the crisis*. Cambridge MA: Center for Advanced Engineering MIT.

Chapter Four

The Change Mandate: A Top-Down/Bottom-Up Partnership

4.1 DELIVERING LASTING IMPROVEMENT

Chapter 1 concluded that delivering future manufacturing success depends on employee engagement and that an efficient organisational design (with a focus on learning) creates a more effective way of working. Chapter 2 introduced the Lean TPM master plan as a route map for delivering engagement, improving effectiveness and efficiency. We also introduced the Oobeya room to act as the central nerve centre for daily and breakthrough management. Chapter 3 set out how to deploy projects to set and manage the accountabilities that secure progress towards the future desired state. This chapter focuses on the three types of improvement techniques needed to make change stick (set out in Figure 4.1). In addition, it sets out how action at all levels of the organisation (senior management, middle management and front line teams) is needed to lock in the changes in management as well as shop floor worker behaviour!

As part of the process, the Lean TPM treasure map supports the transition from ad hoc problem-solving towards *focused* improvement. This also emphasises the evolution of the first line manager behaviours from that of the traditional reactive supervision role to one of team coach as a lean team leader. This is a critical enabler of the future journey from downtime reduction to defect and loss prevention.

Essential tools to deliver lasting improvement

Delivering change to secure lasting improvement requires mastery of:

1. *Frameworks* to make gaps in the current operating model visible and quantify the potential gains from a more effective future operating model. This includes mapping tools, and the asset improvement plan to improve flow, flexibility and focussed improvement. Identifying organisational weaknesses is relatively straightforward.
2. *Change processes* to support visualisation of why change is necessary, build ownership and convert training into new skills. Tools include the use of

71

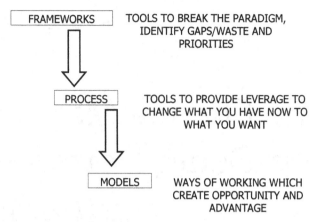

FIGURE 4.1 Tools to deliver lasting improvement.

policy deployment, standard operating procedures, visual management and single point lessons.

3. *Modelling* to shift perspectives and support the adoption of new behaviours and working relationships. In Lean TPM, this is supported by the TRAC (team review and coaching) tool. Modelling and TRAC provide the levers to guide the transition from **away** (reactive) behavioural triggers to **towards** (proactive) triggers.

For example, as breakdowns are systematically reduced, to maintain progress, there is a need to shift attention away from reducing stoppages towards reducing minor quality defects. Minor quality defects are defects that the customer will not notice but which are a barrier to meeting future expectations. The level and nature of minor quality defects can also provide early warning of a process which is losing capability. This transition may sound simple and a logical consequence of any improvement process. The reality is that many organisations flounder and slip back at this stage because they do not refine accountabilities or increase the precision of key performance indicators to maintain a pressure for improvement. Some people may also require convincing that the relentless pursuit of defect reduction is cost-effective. Years of misinformation about the excessive cost of chasing the final 1% of quality has resulted in deep-rooted accepted rules of thumb, which though unproven and incorrect are never challenged. For those of you who hold this limited quality improvement myth close to your hearts, in the real world, customer expectations for quality are going only one way. Not developing that quality capability in advance is a recipe for lost future competitiveness. Also as quality precision is improved, levels of intervention are reduced as do energy costs, productivity and practical batch sizes. Opportunities for low-cost automation increase as does the potential to improve material yield and use thinner materials.

Together these three tools make opportunities for improvement visible, support the adoption of new working practices and reinforce behaviours to revise the operating model to sustain progress. This takes time, care, coordinated effort and must be understood by all in the factory if it is to succeed (Standard & Davis, 1999; Spear, 2010).

As such the major revolutionary changes in the business model may not be believed by employees or necessarily understood even if managers find it completely logical. The delivery of the new model will require the creation of new working relationships – these relationships determine the efficiency and effectiveness of the 'desired future state organisation' (Kurogane, 1993). As a result, the change process will be different for each organisation and one formula cannot suit all businesses and even lean businesses will have followed different change processes to achieve this status (guided by the pressures identified during the policy deployment cycle).

Top-down leadership challenge

The top-down leadership challenge is therefore to design and lead this transformation – to set direction and find the leverage that will help individuals to overcome their natural preference for the status quo and invest energy into learning new skills to engaging with the transformation.

Research into how individuals learn to use new techniques provides some clues as to how to support this learning process (Cummins & Townsend, 1999). Two sets of teams were set similar tasks, one that had classroom training in problem solving and a second that did not. The results suggested that a significant factor influencing team success was not training but confidence. Factors identified by the study that impact confidence were:

- past experience of the same or similar problems;
- the environment and the team's approach to risk.

Training is important, a confident person with appropriate training will achieve more than without. It is application that is most important and that where individuals have limited experience of doing so, support is needed. Confidence depends on the level of trust in management. Figure 4.2 sets out a proven high-performance teamwork development framework which is incorporated in the TRAC audit coaching process. The graphic shows how individual and team development steps are dependent on the progressive evolution of team leader styles. This is in line with the concepts of changes in task behaviour and relationship behaviour described in *Situational Leadership* by Paul Hersey and the *One Minute Manager* by Ken Blanchard. The Lean TPM TRAC audit coaching process builds on this evolution to provide a practical stepwise framework for simultaneous high-performance team leader and team development.

The stepwise framework supports the process of change towards progressively refined models of operation. This sets out a clear ladder of expectation making the search for better ways of working as a core part of everyone's role. Senior management maintains a creative pressure to sustain progress throughout this journey through the processes of:

- setting priorities (consistency of purpose);
- setting standards and supporting delivery (collective discipline),

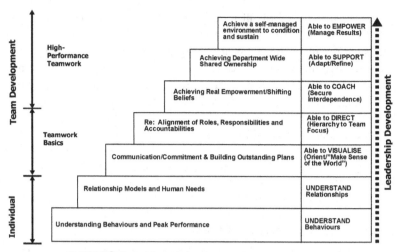

FIGURE 4.2 Leadership/steps to high-team performance.

- Holding people accountable and giving recognition when successful to reinforce the quality of individual learning and the 'right behaviour' sought of the new business design (objective feedback).

As we know, the first principle of 'constancy' is embedded within the policy deployment process. The master plan and TRAC standards provide the team-based processes to set and raise standards. The third element is where management frequently fail. The biggest failing found by researchers and attributed, by teams, to team leaders by over 600 successful teams was a tendency for managers to let people get away with poor performances (Lafasto & Larson, 2002). You may be surprised at this finding – teams ask for discipline! Or importantly the enforcement of standards.

It may seem counter intuitive that teams want discipline and see as unacceptable managers who back down or turn a 'blind eye' to poor behaviours. Allowing ill-discipline, at the team level, creates instability in one of the key elements of the factory's daily management. It also suggests that the training of the team leader or manager has not been totally effective and these matters cannot be treated lightly. When all is said and done, the future of the business and its longevity rests upon everyone working in the right direction. That is why the TRAC model uses behavioural rather than task-based benchmarks. That means that team members will be subject to peer pressure to confirm to expectations as long as progress is supported and reinforced by the recognition process. Team leaders that run counter to this direction, after training, counselling and discipline, cannot be allowed to remain in position. They not only slow progress but serve as focal points to legitimise poor behaviour by others.

At a Lean conference held in London, a delegate asked 'what can a manager do if someone decides they do not want to stick to the rules and take part in a 5S programme' – a team leader in the room had an answer that none of the expert panel came up with. She simply said the manager needed to 'man up' and so long as the employee had been trained and developed and that the 5S workplace discipline programme was core to the company's way of working – then the employee should be taken to the human resource (HR) department via the Trade Union offices. No Trade Union would defend behaviour that undermines a basic safety system of a business and HR would also stand firm with this view. One person cannot refuse and put their team at risk – the consequences of something going wrong and someone getting hurt by the neglect of a single employee do not bear thinking about. If you have done your best – you have trained the individual and they refuse without grounds for refusal then it is not time to back down. Remember teams expect standards to be enforced.

The bottom-up delivery challenge

The *bottom-up role* may be defined as:
- the consistent application of best practice (capability);
- the sharing of lessons learned and reflection upon success and failure (openness and learning together);
- involvement of all stakeholders in the factory area (including trade union representatives, safety representatives and support staff);
- the problem ownership and continuous improvement (aligned goals); the delivery of these roles is dependent on the joint development of:
- a clear compelling future model that shows growth and by default a security from continued employment;
- a practical and supportive change process to deliver the new model that is introduced to allow shop floor personnel to visualise what is expected of them, to develop and implement personal plans of action to respond to what is asked of them;
- the total immersion by management and shop floor in the current milestone or development step to establish new, more productive ways of working.

A good way of confirming the state of this partnership is to listen to the language used throughout the organisation. Do people talk about 'working together to make a difference' or do they blame each other? Are people at meetings positive and proactive or are they defensive and careful to avoid taking actions? Do teams make plans together or plot against each other? It is also interesting to witness whether the bias is towards highlighting past failures and sporadic events or to what can be improved. These indicators reveal the extent of promotion and nurturing that is needed and the involvement of senior management and line management in engaging the workforce in change. If you are cynical, visit a World Class organisation, talk to the people and watch how they react collectively to their performance challenge. That is what a 'World Leading Organisation' sounds like. Most managers think that this can be achieved as a one hit change. In reality, it is achieved through the accumulation of small wins over time. Not an easy task but well worth the entry price and well worth rewarding good behaviour on a daily basis. A simple 'thank you' after highlighting good behaviour goes a very long way. Thanking someone for respecting their co-workers during change, following a scientific method for improvement, managing with humility so that the team can discuss their views and such like are all examples of good practice.

The influential quality guru W Edwards Deming knew well the psychology of promoting a positive attitude to underpin the Total Quality Management framework. The job of every manager is to engage his or her team by creating a win–win partnership to promote effective and exciting change. The role is heightened during the policy deployment cycle when the manager must explain why a certain course of action has been adopted by the entire management team (as a result of the 'catch ball' process).

It is easy to be fooled by the belief that problems are unique to the business or process – one of Deming's bug bears! He refused to accept that businesses are unique and face unique problems. To Deming this was a great way of abdicating responsibility and being delusional. In reality few problems are truly unique, if nothing else, a common thread is that they involve people. Just imagine, for a moment, what could be achieved with this outlook is in place and working across your organisation.

4.2 SUSTAINING THE CHANGE MANDATE

Jack Welch at the head of General Electric (GE) built a step out or market leading growth strategy by adopting a goal of becoming number one or two in their chosen markets. This led to a programme of divestment or acquisition where GE companies could not achieve this through organic growth. Initially successful, the rate of growth eventually began to slow down. After much soul searching it was identified that managers had begun to frame market boundaries so that they could achieve 'number one' or 'number two'. The strategic goal that had once engaged the leaders of organisation with the challenge of 'step out' growth had served its purpose, it was time to move on. The outcome was to redefine chosen markets based on current turnover being 1/10th of that market. With hindsight this outcome may seem obvious but the reality of business is that successful strategies are built and refined over time. It takes true leadership to recognise the need for change and to move away from an approach that has provided success in the past.

No approach can be successful forever, therefore, such change is inevitable. The challenge is recognising the need early enough and taking action whilst there is a choice of futures.

In practical terms, this means being driven by a passion for Business growth rather than managing the status quo. It is the passion for growth that ensures future job security, business viability and competitive advantage.

The graphic below, repeated from Chapter 1, sets out the frequently repeated business turnaround journey (Figure 4.3). The delay in recognising the need to take action and the value of the lost opportunity are leadership hidden losses. Postmortems of companies that have followed this path reveal that in most cases, the decline avoidable. This is evidenced by research that shows over 70% of organisations in the US entering chapter 11 with a restructuring plan survived (Warren and Westbrook 2009). It is reasonable to conclude that had these plans

Results

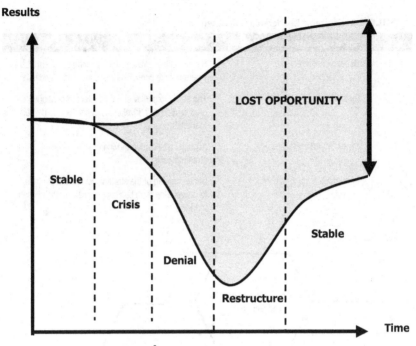

FIGURE 4.3 Leadership – hidden losses.

been actioned earlier, insolvency would have been avoided. Adopting a growth strategy and the planning processes that this requires, is an effective way to mitigate against the risk of missing a fundamental shift in the market.

Imagine that you run an organisation that supplies products to a major automotive manufacturer or food retailer. How would your business cope with the loss of such a contract? Imagine the same business dealing with a significant seasonal fluctuation, caused by exceptional weather. What pressure will that put on the organisation? How quickly would it rise to the challenge? Will departments work together or will they protect their turf?

The foundations for every successful improvement programme are robust business strategies based upon delivering customer value now and in the future (Womack & Jones, 1996; Jackson, 2006). These are also the basis of policy deployment.

Business strategies need to be robust enough to cope with the pressures of likely shifts in the fortunes such as the one above. Scenario planning provides the basis for gaining insight into the consequences of potential shifts in the status quo. The outcome should include the development measures to monitor key trends and the development of contingency plans. That way, if potential scenarios become actual, the warning signs are identified early and because the risks are understood, contingency plans can be activated to minimise their impact.

Table 4.1 sets out the strategic competencies needed to deliver the strategic leadership demonstrated by exemplar organisations.

TABLE 4.1 Strategic Leadership Competencies

	Strategic Competencies	Notes
1	Direction setting	Having a clear strategic intent, being confident about where to take considered risks and show leadership.
2	Building capability	The ability to deploy strategic intent throughout the organisation and develop the capabilities to deliver that intent.
3	Synchronised execution	Aligned organisation moving in step to deliver strategic goals.
4	Performance management	Understanding of the strategic levers and controls to deliver long term business success in the chosen market.

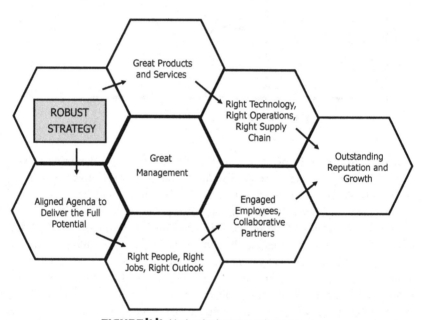

FIGURE 4.4 Underpinning strategic processes.

To master these strategic competencies the change mandate must extend beyond the shop floor to include the strategic processes illustrated in Figure 4.4. The hexagons represent interlocked and parallel management activities and the arrows that the strategic processes that convert intent into practical reality.

There are two paths through the hexagons. A robust strategy is one that delivers outstanding reputation and growth over long periods of time. Making that happen is a process of combining the top path of commercial value development with the lower path of organisational development. Although the deliverable is outstanding reputation and growth, the most valuable outcome is Great Management who are able to deliver winning strategies over time (see Table 4.2 for more details). This is the ultimate goal of any CI process the lower route

TABLE 4.2 Hexagon Map Descriptions: From Strategic Intent to Winning Strategy

Component	Questions
Robust strategy	How to remain successful despite un predictable (but likely) shifts and external pressures
Great management	How to build trust through discipline, consistency and recognition. How to guide the business through the strategic landscape.
Organisational Development Process	
Aligned improvement agenda to deliver the full potential	What is the full potential, what do we need to achieve, what are the transformational milestones to get there and what are the actionable next steps.
Right operation, right jobs right, right people	How to design ways of working and select the right people and develop their full potential.
Engaged employees, collaborative partners	How to create the environment where employees are fulfilled by delivery of business results and partner organisations work on a win–win collaborative basis.
Commercial Value Development Process	
Great products and services	What product and service innovations do we need to lock in profitable customers and disrupt our competitors? How do we make these easy to manufacture and difficult to replicate.
Right technology, right operations, right supply chain	What technology helps us to create low cost, flexible operations to lead current and future the customer agenda. To deliver outstanding return on investment to become the company to invest in.
Winning Strategies	
Outstanding reputation and growth	How can we become the customers' first choice and attract the best people.

focuses on the organisational development and engagement of the workforce with customer focus. The upper route focuses on product and service development which typically includes the capital investment planning, equipment procurement and project delivery processes. The ability to simultaneously manage progress along these two is what characterises the differences between good and great management team capabilities.

Therefore the evidence of great management is a clear and consistent strategic vision capable of aligning personal agendas at all levels behind a common vision. A common strategic vision which militates against the creation of multiple silo management based initiatives which confuse and confound the change agenda.

Understanding the full potential

To support this process of developing great management capability Lean TPM is supported by a 'treasure map' setting out categories of hidden losses and wastes classified against each organisational level (senior management, middle management and front line teams). These are the main policy deployment levels within the organisation which means that the treasure map provides the insight of cause–effect mechanisms to support the design and delivery of a single change agenda.

The treasure map logic supports the primary lean rule of understanding what customers' value and what they are happy to pay for. It also clarifies how each level can release the full potential of the current operation and shape it to better meet future customer needs. As the early steps frequently require little in the way of capital investment to achieve significant performance improvement it can be truly described as a way of 'releasing the hidden factory'.

The Lean TPM treasure map provides a set of lenses to identify constraints to releasing the hidden factory. Initially this can generate a number of questions, providing a good general learning process for managers undertaking the strategy review. Those engaged in the review will often find that their initial assumptions will be challenged. The Lean TPM treasure map includes a set of maturity benchmarks to assess current status and tactics to support progress from zero to hero. The benefits of these tactics can also be converted into potential financial gains using the Lean TPM Loss Model.

This deliberate and iterative process helps managers to refine plans and learn how to view the organisation as an interdependent set of working relationships (Senge, 1993; Liker, 2004).

The strategy review process passes a number of stages which includes the documentation of each iteration so that earlier assumptions can be reviewed when during the later stages, detailed choices become clearer:

1. Issue mapping: Use the two routes set out in the hexagon map. What are the commercial value issues facing the business, where are operational development issues (see Lean TPM Treasure Map).
2. Agree the core focus for business, what is the reason for current and future success.
3. Develop the draft 3- to 5-year vision (linked with the process of policy deployment). Make it one which everyone can get behind. This may require analysis of options against the core focus.
4. Scenario planning: What flexibility do we need the concept to cope with fluctuations in demand (best case–worst case–most likely). Develop scenarios to cover the four main business risk areas of market, cash flow, operations and throughput volume.
5. Agree the recipe for success (include cultural development) and what are the three or four 'must do' ingredients. Identify what is at the heart of future success and the linkages between the 'must do' activities.
6. Use the Lean TPM master plan to set out the transformation milestones and change process (see Chapter 5) and supporting.
7. Apply the policy deployment process set out in Chapter 3 to mobilise the actionable next steps.

Conducting these analyses as a group of cross functional managers may take time but will result in a holistic and better grounded change model (Kurogane,

1993). We will now explore the major wastes to be considered during these management-level discussions and research.

4.3 WHAT DO WE WANT FROM SENIOR MANAGEMENT?

A common heartfelt demand from those involved in continuous improvement is the need for management commitment but what does that really mean? How can you measure it? Remember, what gets measured gets done. The next section sets out the parts of the Lean TPM treasure map which only senior management can deal with.

Wastes in the value chain

Internal analyses and the redesign of the production system is only one part of a much bigger and 'waste laden' supply chain. Most businesses, whether they care to admit it or not, are dependent upon the performance of their suppliers for their own performance. For decades, supply chains have been enshrouded with mistrust and adversarial relationships (Cox, 1996). The approach to suppliers has therefore been to exclude them from anything beyond selling their product and quoting a price. However, the supply chain is also founded upon relationships. Suppliers provide products or services which are not core to the business or which can be made by them more cheaply. For modern firms, traditional supply chain management practices will not support the creation of 'manufacturing-led' competitive advantage and require a similar approach to close working relationship if these important external resources are to be exploited for advantage. Improving interfaces between the organisation and its suppliers affords great opportunities to yield savings in a relatively short space of time. In some cases these can dwarf the savings that could be made by focussing on the internal value stream.

It is strange to think that, for decades, adversarial relationships and deliberate exclusion of suppliers from involvement with the customer has been the main form of relationship for most industrial companies. It has been a default strategy that has not served industry well. Instead resulted in high minimum order quantities, vast amounts of time spent tendering for suppliers, buying from sources in cheap labour economies (but ignoring the true cost of supply) and a lack of synchronisation between forecasts and the actual manufacturing of products. The next sections will explore some of these issues.

Understanding how to exploit and eliminate these wastes can have a tremendous impact on the efficiency of material flow as well as 'open the door' to intercompany collaboration on such issues as the design of new products, stock and cost reductions. A good supply chain does not happen by chance, it is designed. Senior management must therefore focus on the design and refinement of the value stream of suppliers and here it will be found that historic relationships have resulted in six supply chain wastes and practical areas where the value generating capability can be improved. These wastes affect the availability, performance and quality of the materials exchanged. The elimination of these wastes enhances material flow through the external value–supply chain and embodies two of the five main principles of 'lean thinking'.

The hidden losses which only Senior Management can address, how to identify them, their impact and how to reduce them are described in Figure 4.5. This provides a hidden loss treasure map to help identify potential for improvement.

As shown in Table 4.3 these six losses can be categorised as **planned** losses due to inefficient business processes or **unplanned** losses which occur due to the weaknesses in business process design.

This categorisation helps to identify if the route to loss reduction is through improved organisation, improvements to systems resilience or both.

Integration losses

BUSINESS PLANNING

Frequent reactive changes to the production plan indicate that the internal and external links in the value chain are not synchronisation with the flow of customer demand. The more frequent the changes the higher the risk of customer service failure. In most cases, planning rules are never questioned but often include untested assumptions concerning capability to supply. It is typical for a plan to be developed in detail well in advance only to be consistently ignored in the 'heat of battle'. In addition to internal problems and risk to supply these unpredictable swings in order volumes mean that suppliers resort to inventory buffers or engage in expediting work to protect their manufacturing process.

For most supply chains this demand amplification (the variation between what is forecasted and what is actually taken) can be huge resulting in stockpiling by all companies. These stocks add cost and slow inventory turns no matter how good your door-to-door overall equipment effectiveness (OEE) figures. The level of initial plan achievement is a good indication of how well-planning parameters co-ordinate resources in line with true demand and how well the customer business has been in levelling demand to within controlled tolerances with the supplier. A plan should be realistic and achievable. That is realistic in terms of meeting customer expectations and achievable in terms of time and resources available. Poor planning parameters can lead to significant hidden

TABLE 4.3 Understanding Value Chain Wastes

Senior Management Value Chain Losses (S2C)		Integration	Transformation	Customer Value
	Unplanned	Business planning and deployment	Process losses	Unmet future customer value
	Planned	Logistics network	Process optimisation losses	Unmet current customer value
	Goal: effective use of potential resources			

	Loss Category	What is it	Why is it bad	How can we find it	How can we reduce it
Integration	Business Planning (Long/medium term)	Planning assumptions which do not mirror actual results, forecast error levels, reactive planning activities.	A poorly tuned business planning process can reinforce fire fighting, and 'just in case' behaviours. This constrains the implementation of strategic plans and cultural change.	Measure plan adherence and level of performance improvement, Assess business processes against Business Excellence model criteria or similar.	Clarify strategic vision and medium term "must do" goals. Deploy priorities through accountability setting and change management process
	Logistics Network	Under utilised assets, distribution network or inventory. Weak alliances/ outsourcing tactics	Surplus resources/stock adds to fixed costs and masks hidden losses. Alliances which are not developing towards a win/win basis are at risk/in decline	Value map supply chain and inventory movement profile, Identify high transaction costs and material/service cause/effect mechanisms.	Improve logistics/operations strategy. Target non value adding processes and total cost of ownership. Automate document workflows.
Transformation losses	Process Design Losses	Technology which is difficult to use, look after or is wasteful in terms of resources	Poor design effectiveness increases operational intervention/skill levels, the risk of defects/HSE issues and reduces ROI	Analysis of ease of use, process characterisation, variability in defect levels and energy/tooling usage.	Early Management of product and capital projects to deliver flawless operation from day 1.
	Process Optimisation Losses	Inconsistent process quality or resource needs.	Lack of control of chronic losses results in higher levels of costs and quality defects	Stable or declining improvement trends	Use Quality Maintenance/6 Sigma techniques to establish and deliver optimum conditions
Customer	Unmet Current Customer Value	Ability to develop and launch successful new products and services	The ability to create new value is essential to business growth	Slow growth in profits from new products, stable or declining time to market,	Excitement feature VOC innovation, Product life cycle management
	Unmet future Customer Value	Lost sales, weak customer loyalty or reducing market share	Indicates lack of understanding of market shifts and potential loss of competitiveness	Share of Market and Customer business, VOC survey	Total cost of ownership analysis, Basic and Performance VOC innovation

FIGURE 4.5 Dealing with supply chain wastes.

waste throughout the organisation as people engage in expediting or sorting out quality defects before they affect the internal OEE calculations (Nakajima, 1988). Reducing the chaos of poor planning management has many benefits not least in ensuring that processes are available to produce what the customer wants in the smallest batch sizes possible (maintain a low production cost). Demand amplification and planning chaos also means product costing information is based on parameters which do not reflect reality as production consumes overtime, batches are cut short and all manner of intervention is needed just to get products out of the door.

Demand amplification also means, although improvements are made at the door-to-door and equipment levels, this potential is not translated into real benefits. Until improvement activities reduce inventory little has been achieved. Effective businesses use these improvements to reduce inventory and to lower costs of operations (Standard & Davis, 1999). Staggering improvements in changeover time which do not reduce total inventories deliver little real business benefit and consume a lot of employee good will. This important point also has implications for generations of engineers who have sourced the quickest and fastest processing machinery when given a capital expenditure budget to replace existing machinery. Buying the 'state-of-the-art' machinery, to replace existing assets, must result in a redefinition of the stocks needed to support the business. If this does not happen then the potential benefit of the quick cycle time of the new machine is lost in a mass of inventory. In reality, value added of factory operations has worsened because excess inventory has not been removed.

This form of investigation also applies to other planning parameters. For example an investigation into delivery failures in a chemical plant revealed how planning rules based on large batch sizes meant that despite spare capacity, opportunist orders won by salesmen were impacting on orders to customers who had given long lead times.

LOGISTICS NETWORK

Often slow moving or specialist products are a variant of another product possibly a foreign language version or pack size. Storing intermediate stock or assemblies and making such products to order can provide the flexibility to deal with such difficult to forecast demand items. In this example, the intermediate stock is the decoupling point between supply and demand – this is the point where resources are focused to accommodate demand. Lean TPM seeks to move the supply decoupling point as far back towards the raw material supply as possible. As we know, if demand can be met by making to order within the customer order lead time this will provide the lowest cost supply option. Typical flow management losses include spare capacity and excess inventory:

Spare capacity in excess of customer demand is not necessarily a bad thing where it provides flexibility to protect against seasonal or customer distress purchases. These events are often predictable and can be controlled and for many companies. Even businesses affected by the weather temperature or sun shine tend to have product characteristics (shelf life) that allow these events to be accommodated without causing chaotic production. The ability to flex capacity can, even under these conditions, be an important component in maintaining customer loyalty and providing customer satisfaction. Figure 4.6 below illustrates how year on year improvements in OEE doubled capacity over 5 years through a focussed improvement programme. The decision to improve make ready and raise quality standards was taken at a senior level to support a strategy of business growth. As a result resources were targeted at the most attractive product areas and accountabilities across all functions were aligned through the policy deployment process. This increased the capacity of the logistics network and the number of customers it could supply. The benefit of delivering the forecast business growth at the improved OEE level included profit margins increased by over 200%. Had the increased capacity not been required the gain would have been around 40%. So by embarking on this strategy the management team took a risk but the gains included a more engaged work force and increased logistics resilience. Even if this had not generated additional profit, the journey would have been more than worthwhile.

In defining a change mandate, by looking at capacity issues and the predicted benefits of a Lean TPM approach, the ability to sell manufacturing capacity is important. The most effective way of engaging the workforce behind the improvement agenda is to follow a growth strategy and relating improvement to more sales at relatively little additional effort. For the workforce, growth is a way of avoiding problems with what to do with labour displaced by improvement activity. Without this approach, improvement activities will falter as individual workers correlate improvements with the loss of jobs for co-workers. The

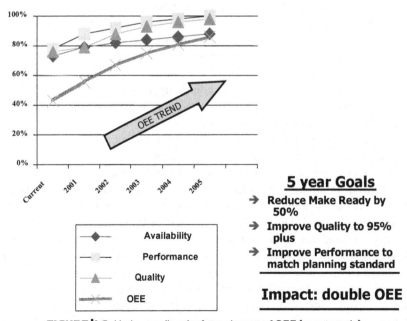

5 year Goals

→ **Reduce Make Ready by 50%**

→ **Improve Quality to 95% plus**

→ **Improve Performance to match planning standard**

Impact: double OEE

FIGURE 4.6 Understanding the future impact of OEE (a case study).

win–win proposition is to achieve a growth rate which is equal to or better than the rate of improvement. In most industries the biggest single factor influencing profitability is order volume so this also makes commercial sense.

Surplus inventory and stock is often a result of trying to plan too far ahead and in too much detail (Womack & Jones, 1996). In particular:

- the longer the planning timescale the more difficult it is to predict what will happen;
- the more detailed the plan, the less willing people are to change it to reflect reality.

To lean businesses inventory is classified as an 'evil cost' – it is sleeping money that is not earning a return. As such, lean businesses employ strictly controlled kanban stocks to limit this exposure to idle money and pay attention to maintaining flow. Also asset care activities are used by lean businesses to ensure assets do not fail and require safety stocks to cover for this failure. These stocks are in effect an apology for poor maintenance. Low inventories make sense both in terms of 'business logic' and reinforcing the importance of asset care and development of diagnostic skills. The other reason for deliberately lowering safety stocks is to force improvement and allow abnormalities to be detected quicker and action taken by the appropriate operational personnel.

The planning process does not have to be complex. There are only two planning parameters to manage:

- how much to produce;
- when to produce.

minimum coordination is achieved when:

- medium term planning monitors demand trends to assure that outline resource–capacity allocation rules can match true demand;
- detailed planning is delegated as far down the organisation as possible.

In this way, decisions can be made as late as possible rather than attempting to anticipate in advance day-to-day conditions such as labour availability or last minute changes in customer demand. Experience shows that in some industries it is possible to use simple rules to coordinate up to 80% of the production volumes. For most businesses demand profiles can be broken into three categories of product manufactured at the business:

- *Runners* (5% of products providing 50% of revenues);
- *Repeaters* (15% of products providing 30% of revenues);
- *Strangers* (80% of products providing 20% of revenues).

Demand for products classified as runners is fairly predictable and finished goods safety stock levels can be low. The decision on 'how many' or 'when to make' can be controlled by simple '2 bin system' or kanban where work is pulled by actual demand. The variation in demand for repeaters is higher so that it is only possible to predict overall demand. These are the items where some batching of production will be necessary. They can be made at any time capacity is available.

The variation of month-to-month individual stranger demand is very high making it impossible to forecast accurately. It is generally possible to forecast demand for groups of stranger items and from this how much capacity they may require. So although products with demand profiles classified as Strangers can only be produced to order, it is possible reserve capacity for them as a collective group. This capacity can then be allocated as orders arrive keeping lead times to a minimum. If stranger levels are lower than forecast, the capacity reserved for them can be reallocated to make runner or repeater products.

The net result is that flow management tactics should reflect the predictable patterns of demand profile. Ideally, these should leave the choice of what to manufacture as late as possible by using simple capacity allocation rules which can be managed as close to the point of manufacture as possible. Under these conditions, opportunities to use available flexibility will be encouraged as will activities to further increase flexibility to changes in demand.

Transformational losses

Process design to deliver high-flow, low-defect, low-cost products, processes need to be intrinsically safe, reliable, easy to operate and maintain. It is common to find that the opposite is true. Equipment is difficult to operate and maintain.

Another design waste is that of overspecification of equipment. This has a number of costs associated to it not least that of increasing the amount of losses during processing. For example, an oven that is bigger than needed will take longer to heat up, cost more to run and maintain and usually cost more to buy. In addition functionality which is not required adds to the complexity of the design-making problem solving and optimisation more difficult. It is also important to identify where potentially useful features are not used because they are in a poor state of repair. These issues go right to the heart of the 'automation debate' and the 'right sizing' of productive assets to do the job expected of them using capable technology. All too often, neglect and even engineering experimentation has resulted in an asset base that is in need of review. It is interesting to note that many 'world class' organisations have basic policies which involve precluding experimental and unproven technology from the workplace as these assets are, by nature, problematic and risky. Also many of the 'world class' organisations, such as Toyota, also do not automate a process that has not been perfected by rounds of problem-solving and engineering assistance (Monden, 1983; Liker, 2004).

With the relentless pressure from customers for new products, the ability to deliver new products and equipment that achieve their full potential is an important strength. This will therefore have a major impact on the growth strategy of the business. If you want to predict how well your future technology will support competitive advantage, review the current technology to provide:

- An assessment of current technology–product platforms and their fitness for purpose. This should highlight strengths, weaknesses and where current functionality is not being used.
- An assessment of capital project delivery performance against early management benchmarks set out in Chapter 6 to identify barriers to flawless operation from day 1 for new products–services and assets.

- Clarification of the Commercial, R&D and Operations accountabilities for improving the project delivery process.

TECHNOLOGY AND ADDED VALUE OPTIMISATION

The Lean TPM Journey passes through the stages of 'zero breakdowns' towards 'zero defects' and, as organisational capability improves, nonvalue adding activities are removed. This is a key enabler of progress towards achievement of market–customer leading performance.

Transportation and motion losses are common wastes where technology has not been optimised. Customers are not willing to pay for the fact you move products over two miles within your factory before it is packed and ready for shipment. Travel distances increase the likelihood of damage and 'world class' organisations seek ways of lowering travel distances to achieve much better just in time performance.

A poor layout design will add miles and hours of unnecessary costs over the lifetime of the product and of the factory. Even process plants that have engaged with Lean TPM are working to the concept of 'pipe-less' plants where buffers between processes are minimised. Pipes are a hidden form of transportation that have their own difficulties including the phenomenon of 'hammer rash', which is the technique used to un-block pipes by hitting them with rubber mallets to increase flow. 'Pipe-less' factory designs, for process plants, make it possible to produce smaller quantities economically and to reduce capital costs, plant footprint, new plant delivery lead times and improve return on investment. This issue, especially making visible this form of waste, is central to the technique of value stream mapping, which will be explored later.

The solution to these problems lies in the development of new relationship and alliance between the marketing department, maintenance, operations and R&D functions to deliver an asset base and production system that is sized to deliver what customers value. This is one of the fundamental drivers of early management (see Chapter 6) and quality function deployment (QFD, Cohen, 1995; Ficalora & Cohen, 2006). QFD incrementally drills down to define in more detail the features highlighted by the Voice of the Customer profiling. This process converts desired features into actual specifications and operating conditions for the asset to produce them (Bicheno, 2000).

Customer value

Unmet Customer Needs refers to both consumer products and services like vendor managed inventory and line side deliveries. Research by Christensen (2002) suggests that targeting unmet customer needs resulting in disruptive innovation, produces products and services with a one in three chance of success compared to a 1 in 15 chance of success for 'me too' products. The research also highlighted that when leading companies were toppled, it was almost when disruptive technologies emerged. Although this research focussed on technology-based products, the findings are relevant to many other industries.

Understanding the 'voice of the customer' is, therefore, an essential capability for any manufacturing business. Failing to understand the purpose for which the product is being purchased and believing that your products can remain unchanged for many years is no longer credible. The car industry is a good example. About two decades ago, the materials and components used would have been relatively standard and unchanged. Today the modern car industry upgrades these items every 6 months to allow improvements to be assembled into the vehicle. Even materials such as steel have been subject to improvements in bodywork warranty and anticorrosion performance. In short, it is easy to miss and not meet customer needs.

The importance of the service–product package should not be underestimated. In retailing one of the areas of increased margin in recent years has been the offer of extended warrantees. Finance deals are an increasing source of revenue for automotive manufacturers. In business-to-business industries, understanding the customers total cost of ownership analysis can provide ideas for reducing costs to the customer. This also has the potential to increase supplier revenues because the product–service offering is more valuable to the customer. It may even result in different ways of charging for the product–service. For example, a supplier of oil and gas platforms undertook a project to build a platform for a major customer where part of their revenues were paid based on the barrels per day output from the platform. They were able to share the gains with the operator of improving the design in ways which reduced capital cost and increased output.

Review the past performance of your business against the four areas set out in Table 4.4. What has been tried, what was successful, where is the future potential? Further analyses should consider 'What would a competitor need to do to win business from this company?' This analysis, at least, provides clues as to where to investigate to create a portfolio of product and service offerings to satisfy the modern customer.

These analyses typically illustrate how collaboration within your supply chain provides an effective countermeasure to a stagnant marketing mix of products and opens new opportunities to engage suppliers with the customer market for mutual gains.

The targeting of 'On time in full' product supply is in theory the goal of every business, yet many accept a trade-off between cost and service levels. Increased reliability and flow potential will make it possible to reduce lead times

TABLE 4.4 Understanding Customer Needs and the Impact of Technology

		Technology	
		Current	New
Customer needs	Unmet current customer value	Replacement products	Disruptive products
	Unmet future customer value	Incremental products	Customer leading products

and improve forecast accuracy thereby eliminating unnecessary stocks. This has an impact on customer loyalty. Research into 'why customers change their loyalties to a certain manufacturer?' suggests that around 10–15% of customers change suppliers due to changes in circumstances and around 30–60% routinely assess supplier performance. Understanding the losses associated with delivery performance is therefore important if senior managers are to understand what drives the customer and what should drive the supply chain in order to effectively focus improvement activities.

4.4 WHAT DO WE WANT FROM MIDDLE–FIRST LINE MANAGEMENT

In any change programme, it is the middle–first line management level that dictates the rate of change and is also the most problematic to deal with. This is the level of the organisation where the strategic intent meets the white heat of shop floor reality. Often unintentionally this layer of management can hinder change by translating top-down input by filtering out what is seen as impractical rather than embracing the challenge.

The same can happen when translating bottom-up input to remove unrealistic requests that will not be considered. This is also the level of the business within which politics and 'guerrilla warfare' tactics have traditionally been used in power struggled between departments (Rich, 1999). Of note is the power struggled between operations and marketing (Brown, 1996). This is ironic in that both these two departments rely upon each other but more often than not this tension still exists in modern manufacturing firms. The problem between these two departments rests upon the issue of 'customer service'.

For marketing, the operations are a drawback and constraint that prevents the business from moving forward. For operations staff, 'marketing personnel' are often regarded as 'know nothing' salesmen who have little appreciation for the engineering input, capacity management, and other issues required to deliver products against the contractual requirements set by the sales and marketing staff themselves.

This relationship, however, lies at the crux of 'world class' manufacturers. Such organisations have managed to create a superb brand for the quality of product supplied and to have correctly and effectively harnessed the manufacturing operations of the firm as a means of competitive advantage. They have achieved this through breaking down the internal boundaries and improving communications to achieve a mutual respect and understanding across functional boundaries.

Door-to-door losses

The improvement focus of first line managers (FLMs) is door-to-door losses. These loss categories highlight problems due to gaps in top-down policy and difficulties in shop floor operation. As such they provide a framework for managing upwards as well as a one for coaching FLMs to develop a wider appreciation of the business.

Door-to-door losses are categorised under the headings of:

- preparation;
- coordination;
- adherence.

These forms of loss represent targets for improvement and gaps in the fabric of management. They help to unpick the complex cause and effect relationships, which result in lower true customer service and business effectiveness. The outputs from this analysis help to direct resources towards the most important improvement potential (Figure 4.7).

As shown in Table 4.5 these six losses can be categorised as **planned** losses due to inefficient business processes or **unplanned** losses which occur due to the weaknesses in business process design. This categorisation helps to identify if the route to loss reduction is through improved organisation, improvements to problem prevention protocols or both.

Preparation losses

Preparation concerns those tasks carried out to support the production process. This includes inbound logistics and administration as well as maintenance and cleaning activities. Often such activities are ignored despite their potential for improvement.

Inbound logistics losses can occur when resources are not available to begin production and the reasons can be many and various. Some of the most common reasons include goods inward problems, poor shift control and poor utilisation of engineers even inconsistent or irregular production scheduling. There are many reasons why the schedule itself could be irregular and exploration of these issues often finds many problems associated with 'dead data' in the planning system. Other issues include not scheduling resources because of unexpected interventions made by the sales department to reprioritise orders and accommodate sales of products within the stated lead time of the company. These issues will also be reflected in low schedule adherence. The solution is the planning process and to focus upon creating stability for the manufacturing process in terms of getting materials available and communicating manufacturing schedules such that labour is available.

TABLE 4.5 Door-to-Door Losses

First Line Manager (D2D)		Preparation Losses	Coordination Losses	Adherence Losses
	Unplanned	Supply–resourcing failures	Expediting losses	Schedule adherence
	Planned	Activities to prepare for routine production	Processing losses	Direct cost of quality
Goal: Effective use of internal resources				

	Category	What is it	Why is it bad	How can we find it	How can we reduce it
Preparation	Inbound logistics	Resources materials or instructions fail to reach the first process as planned	Can result in wasted effort and increase the risk of customer service failure.	Monitor delays to campaign start times. Analysis of inbound logistics	Improve information flows, target total cost of supply, supplier managed stocks
Preparation	Ancillary operations	Additional activities needed to prepare for the next campaign, includes cleaning, maintenance or pre production inspection	These resources can be hidden in fixed costs and not subject to CI. Their impact on productivity can also be undervalued leading to under investment.	Value mapping of door to door internal supply chain. Day in the life simulation, Analysis of fixed costs	Process reengineering to eliminate, combine, simplify ancillary operations.
Coordination losses	Expediting	Unplanned delays in feeder to fed processes or additional activities to expedite production.	Interrupted process flow means higher level of intervention, coordination and space requirements raising fixed costs	Line studies, value mapping, labour productivity KPI's, WIP levels, Resource usage.	Synchronous production techniques, e.g kanban, Improve process reliability
Coordination losses	Process flow	Activities such as transportation, double handing or in process inspection delays	Extends production lead times, adds to WIP and intervention levels. Can result in the need to expedite backlogs	Value mapping	Improve product flow using feeder to fed processes and pipeless process plants
Adherence	Schedule adherence	Inability to produce and deliver as planned. Includes overproduction	Results in plans with inbuilt buffers to allow for 'unforeseen' but avoidable problems.	Areas requiring routine expediting, low correlation between planning standards and actual time taken	Review basis of planning standards, simplify scheduling and delegate short-term planning activities.
Adherence	Direct cost of quality	Accepted resource waste/material give away both formal and informal	Can hide wasteful practices or potential for improvement.	Target total yield rather than performance vs cost standard	Product and tooling design. Process optimisation.

FIGURE 4.7 Dealing with door-to-door losses.

Ancillary operations includes operators placed on the line to watch for falling products or sortation (and assembly) line processes where the level of work is not balanced. These 'non-jobs' become accepted elements of the work and often involve operators fetching chairs to sit on while they watch the machines for which they are responsible. Such waiting can be reduced through multi-process operations using walking operators who tend to many machines and this is a well known technique employed by lean organisations. With process optimisation, the need for such watching and interventions to cover for material flow failures can be eliminated. Such tasks are boring and often dangerous. Their removal improves safety and quality as well as productivity. Other examples are maintenance, product testing and cleaning. These are essential activities but like set up and adjustment time, they should be subject to review to reduce the time taken whilst maintaining quality standard. Some carry out such tasks at weekends or on night shift and, although this might seem a reasonable tactic, it is often expensive (adding 'unsociable shift premiums', removing the opportunity for learning and problem-solving between teams).

The solutions to these problems include the need to re-industrial engineer processes, to understand the costs associated with the losses and also to engage problem-solving activities with the various factory teams to highlight these wastes in an attempt to eliminate or reduce them.

Coordination losses

EXPEDITING

In many industries, equipment is not the bottleneck resource it is people and their lack of skills. Here the availability of labour or specific skills can be the difference between profit and loss. Reducing and eliminating coordination needs to result in a self-managed system and part of this process is the removal of nonvalue adding activities and sources of unplanned interventions. The need to expedite jobs is evidence of a poorly coordinated process.

PROCESS FLOW

Ideally co-ordination within the value generating process autonomous or self-managed systems supported by visual management set out the game plan, make early completions–late finishes visible at a glance and support the cadence of the workflow. There are many potentially self-managed internal coordination approaches including:

- a total pull system (where product variety allows);
- only scheduling the real bottlenecks in the factory; and
- the use of finished goods stock levels as the trigger to launch materials to the manufacturing system.

The need for management intervention prior to deciding what to make next or waiting for sign off by quality control is an indication of poor job or workflow design and a lack of process reviews to reduce or eliminate these sources of failures. This applies equally to tasks such as maintenance, production or project approval and again a re-engineering approach, using cross functional and operations teams, is needed to combat these issues.

Adherence

If 'door to door' processes are under control, they should lead to easy adherence to planning rules and material–resource usage standards.

Schedule adherence concerns the detailed schedule covering the internal supplier–customer work flows as well as the delivery of end customer orders. This is a major loss area for most manufacturing businesses and is the source of many business failures. Instability in the basic schedule (the information that triggers production) is costly and causes products to be launched too early, in the wrong sequence or too late for the needs of the customer. Furthermore, when order levels are volatile and the schedule unstable, some will 'best guess' the schedule and inflate requirements so that orders to suppliers bear little resemblance to what is being sold. The inevitable outcome is that the business has stocks of what are not selling and no stocks of what is – this is not good for cash flow.

Direct cost of quality covers the material and resource waste are reviewed against a zero base to scrutinise planned or standard waste allowances such as material waste. Typically, material costs are three times that of labour and should be subject to a corresponding level of focus.

This can include yield or material waste, effluent disposal, cutting fluids, packaging, tooling, water usage and other high-cost consumables that are traditionally considered 'overheads' but should be treated as direct costs. This is worthy of a major review by a multidisciplinary management team including sales, marketing, planning and operations management. This is particularly useful where batch sizes are variable and, on certain jobs, the company could be losing money. This information is also relevant to the sales and marketing department who, armed with this knowledge may be able to negotiate a better sales price for the difficult–costly to make products. Cost deployment is discussed in more detail in Chapter 6.

4.5 CALCULATING DOOR TO DOOR OEE

Door to door can be measured using a variant of OEE. Remember effectiveness is a measure of how well you do against the full potential. The losses, against that full potential at a door-to-door level, help us to assess if the loss was due to problems with preparation, coordination or adherence. We can still measure OEE at an equipment level to gain an insight into how much of the coordination loss was due to equipment issues but that is dealt with at the next level down. At this Middle–First line Manager level we are interested in how well the equipment was used to extract value from every minute it is planned to operate. The OEE has three components and is calculated as follows (Table 4.6):

TABLE 4.6 The OEE Calculation

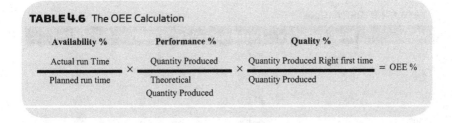

Availability %		Performance %		Quality %	
$\dfrac{\text{Actual run Time}}{\text{Planned run time}}$	×	$\dfrac{\text{Quantity Produced}}{\text{Theoretical Quantity Produced}}$	×	$\dfrac{\text{Quantity Produced Right first time}}{\text{Quantity Produced}}$	= OEE %

TABLE 4.7 Exploring the OEE Calculation

	Description	
A	Planned run time	20 h
B	Set up	1 h
C	Breakdowns	1 h
D	Actual run time is A − (B + C)	18 h
E	Theoretical rate per hour	200
F	Theoretical production when running E × D	4000 units
G	Actual production from production sheet	3000 units
H	Quality rework	100
I	Right first time production G-H	2900

The simplest way to explain the advantages of the measure is to use an example. Assuming the following production statistics (Table 4.7):

The OEE calculation would be as follows:

$$\frac{18h}{20 \text{ Hrs}} \%\times \frac{3000}{3600} \%\times \frac{2900}{3000} \%= \text{OEE\%}$$

This equates to an OEE of 73% (as shown below).

$$\frac{90}{} \%\times \frac{83.3}{} \%\times \frac{96.7}{} \%= 73\%$$

Even though the individual components of the calculation are reasonably high, the OEE itself shows plenty of room for improvement but where should you start? Naturally we want the process to be available when we want it, to run a maximum performance rate and produce 100% right first time quality. Each component of the OEE is linked to two areas of loss. Each loss is a different

TABLE 4.8 OEE Analysis

Week	Availability% x	Performance% x	Quality%	=OEE%
1	90%	83.3%	96.7%	73%
2	85%	85.3%	96.6%	70%
3	95%	81.6%	96.8%	75%
Average	90%	81.6%	96.7%	73%
Best of the best	95%	83.4%	96.8%	78%
Gain	5.0%	1.9%	0.1%	5%

sort of problem so by identifying the priority loss area, you also identify the techniques to reduce it. A further sophistication of the approach is the use of the best of the best targeting. If the OEE is calculated over three or more weeks, the individual results will fluctuate (Table 4.8).

This gain from 73% to 78% provides a realistic and achievable 1-year improvement target for the process. Not only that but the highest area of fluctuation is usually a good place to focus on. In this way the regular recording of losses provides an insight into the shop floor reality to identify weaknesses and monitor improvement trends. The OEE measure is highly punishing but a great way to focus improvement attention and also as a marketing technique to promote the change mandate within the factory. The OEE value for an asset or linkage of assets as a whole is a great way of gaining stability by focusing improvement activities at each stage in the production process with the outcome that each percentage improvement in OEE performance increases the flow of products through the factory. The OEE figure is therefore most commercially important when applied to the bottleneck process and all operations after that asset.

Designing the OEE measurement approach

The initial applications of TPM during pilot programmes or in jobbing shops the OEE measure is typically shown as 'individual machine effectiveness' but in many cases the measure is most useful when applied at a cell or production line level. Here it is important to design the application of OEE carefully so that it provides a reliable measure of effectiveness which is does not suffer from statistical noise or fluctuation due to external factors. It is worth remembering that 'effectiveness' is a measure of how well we achieved what we planned to do.

A well-designed OEE measure should support the process of learning from experience and identifying what stopped achievement of the plan and dealing with the reasons why things do not go according to plan.

Figure 4.8 illustrates a production process where Cell W is routinely fed by cell Z and occasionally fed by Cells X and Y. Initially the company calculated

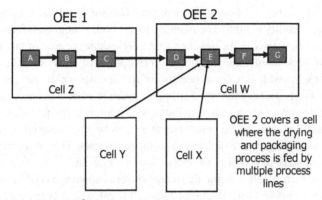

FIGURE 4.8 Designing the OEE measurement system.

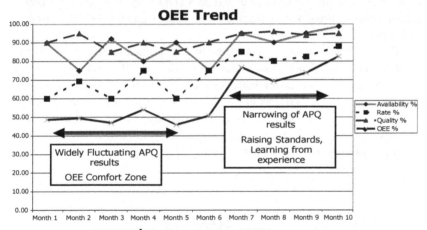

FIGURE 4.9 Characteristics of OEE improvement.

the OEE across cells Z and W as a single flow. This resulted in fluctuations in OEE. In the weeks where cell X–Y fed into cell W sometimes more than 100% performance was recorded. Creating separating measures for Cell Z and Cell W provided far more meaningful results.

What is important is the OEE trend, rather than absolute OEE levels. It should always be improving. It is easy to fudge OEE measures so that they look good. This is one of the reasons why it is meaningless to compare OEE results on different items of plant (such as a heat treatment furnace and a metal press). As long as the OEE information is measured on a consistent basis, this provides meaningful management information to direct and confirm progress towards waste elimination. Figure 4.9 illustrates how the OEE trend behaves as equipment progresses from an out of control to an in control condition. Over the first 6 months of the chart the OEE trend shows little overall change. Note how the Availability and Performance curves cross over. When availability is down, performance is up. In this case it was due to increased pressure to produce output targets following time lost due to breakdowns (a fairly common industry pattern) but, in theory the peak performance should also be possible when the availability is high. The inability to get both should therefore prompt investigation especially as the quality rate is fairly stable because of measures to assure quality. Secondly, the analysis will prompt managers to investigate why, on occasions, speed losses ensue because of quality assurance routines. In this way the OEE measure is a means of telling us much of what is wrong as it is also a quantifiable way of seeing positive impacts of improvement activities.

In the last 4 months, as the improvement process begins to take hold, there is a noticeable narrowing of availability and performance rates. The two curves begin to move in the same direction and with less 'swing' or deviation. This indicates that problems are being dealt with and that learning from experience is taking place.

The 'door-to-door OEE' (D2D) of each cell is calculated by treating the cell as a 'black box' here the OEE of the bottleneck will equate to the OEE of the cell. Even where the OEE is calculated on a D2D basis, it is still important to record losses incurred at each machine. Recording should include door-to-door and floor-to-floor loss categories.

This is because on the shop floor, the most important measures of improvement are the level of losses rather than the OEE itself. OEE measures what we have achieved, losses measure what we need to do better at. Figure 4.10 illustrates the level of losses occurring on a month-by-month basis. The fluctuation shows that despite the OEE becoming more stable in the last 4 months, speed and quality losses are clearly not under control. As will be discussed in Chapter 5, the first priority is to get in control.

Monthly Losses

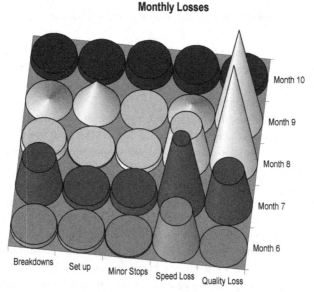

FIGURE 4.10 Targeting loss functions.

4.6 WHAT DO WE WANT FROM FRONT LINE (SELF-MANAGED) TEAMS

The classic six losses, monitored by the OEE measure, are well documented. These are summarised in Table 4.9. These losses are categorised by asset availability, performance and quality issues. Each of these being subdivided under planned and unplanned loss headings.

Breakdown losses

A breakdown is a sporadic failure (such as a belt snapped or a shaft sheared) and these are different to minor asset stops. It is the issue of breakdowns that remains in the minds of operations staff although it is not always the case that these are the largest losses to production when accumulated in an OEE analysis over time. Chapter 5 sets out how to achieve zero breakdowns.

Set up and adjustment

Time spent setting up and/or adjusting equipment is time which cannot be spent producing and so this is considered a loss. Ad hoc running adjustments are also included under this heading. Examples of this include:
- not getting the set up conditions right for the product at the beginning;
- 'fine tuning' by operators and techniques to maintain control of the machine;
- frequent adjustment of materials fed to the machine such as the management of reels in the production of electronic circuit boards.

Set-up losses can be huge and are often ignored by operations staff who have simply grown to live with long times during machine changeover and ramp up. World leading plants achieve zero interventions or no touch production and have through sustained efforts minimised this form of downtime.

The approach to reducing losses includes an analysis of methods, by operations and technical staff, to minimise the equipment downtime using quick changeover and SMED techniques. SMED stands for **S**ingle **M**inute **E**xchange of **D**ies. A procedure used to reduce the machine changeover to 9 min or less (single minutes! Double minutes would be 10+). The ultimate aim is to achieve

TABLE 4.9 Equipment Level Losses

Equipment (F2F)		Availability	Performance	Quality
	Unplanned	Breakdowns	Idling and minor stoppages	Defects and rework
	Planned	Set up and adjustment	Running at reduced speed	Start up losses
	Goal: Effective use of assets (lean operations)			

'one touch changeovers' where the operator simply presses a button and the product is changed. The goal is to achieve a slick 'Formula One' motor racing 'pit stop' when carrying out a changeover. For industries, such as glass making where the setting of a tool can take an hour but ramp up of glass flow can take 5 h, this form of improvement programme offers a tremendous financial, capacity and customer service reward to the firm. From a 'door-to-door' basis, cleaning activities and material co-ordination should be subjected to the same analysis even though, in many industries, cleaning is considered a 'low-grade' task and is consequentially ignored. The key to reducing cleaning times is to reduce the scattering of dust and dirt which in turn extends the workable life of rotating equipment (lower wear) creating a double benefit for the business and material flow performance.

Idling and minor stops

The term 'minor stops' does not necessarily refer to the length of the stoppage but is a means of classifying all those stoppages that often go unnoticed (until measured) but whose cumulative effect can be enormous and drastically reduce operating time (and increase worker frustrations). At an automotive manufacturer a fire caused by the build up of heat due to the density of spot welding tips in a particular body section and the team had learned how to put the fire out quickly and restart the machine. They treated it in a minor incident despite its potential severity. At a cement plant, blockages could stop the plant for half a shift and the response was to clean out the pipes and restart. In each case the problem had not been solved it had not gone away. This is a significant difference to a breakdown. With minor stops the first major hurdle to solving the problem is believing that it can be solved and breaking customary practice (remember the 'hammer rash' problems mentioned previously). Minor stops are highly costly to any manufacturing business; they are annoying to the operations staff and an irritant to maintainers (who are often unaware of these events). The solution is to make these problems visible by correctly documenting these failures and applying problem-solving activities or to engage devices which sound an alarm when they stop which is used to prompt response.

Reduced speed losses

There is often heated debate about how fast the process can run and what standard to use for performance rate calculations. Typically, however, reduced speed losses often occur as a pragmatics tactic to reduce quality defects. That is reduced speed losses generally mask latent quality defects. It can take time and effort to reduce these losses but these losses should always be made visible. For that reason we recommend that the theoretical speed used in the OEE measurement is the highest potential speed. This will produce a lower OEE result but the potential will be clear. These losses can also be calculated and equated into financial losses and cost of lost production. For many businesses this is one of the most difficult losses to correct. This is, as most world class companies will confirm this is the stage that demands the most skills from the engineers and operator teams. This is also the loss that results in some of the most spectacular innovations which can be incorporated in the purchase specifications of the next generations of productive assets.

Start up losses

These losses occur at the start up of production assets and this is typically the major improvement area for process plants, particularly where the process chain is fairly simple such as oil extraction. This loss is also influenced by the set-up activity and in many cases, where plant is left idle for periods of time, the quality of cleaning after the close down can have a major impact on the time taken to get the asset–product back to specification.

Quality defect and rework

For most organisations quality defect reduction presents a major opportunity to improve productivity and reduce costs. The true cost of poor quality is typically huge even though most defects can be prevented using focused engineering and operator efforts. The goals of any improvement team must therefore include zero defects and a determination to waste nothing in servicing the customer. This sounds a perfectly logical objective but even after decades of promoting quality management most customers would be horrified at the extent of reworking and scrap. Such mistakes have to be paid for and the person that does so is the customer. Poor quality performance also has a vicious circle and tends to result in increasing production batch sizes, paying operators to rework products and also using too much bottleneck capacity (or using it twice) to make something that should have sailed through the production process.

However, quality losses remain one of the least understood areas of loss and involves, not the measure of good quality to the customer, but the right first time quality value. In one car plant when this measure was applied the 'real' quality was measured at around 30% for the process. The aim of this measure is not to embarrass management but to illustrate the hidden loss and provide recognition as it was reduced by improvement efforts. In the process industry where rework and blending of out of spec materials is often built into the process, the quality % should reflect this. In a cement plant before this rework measure was introduced it could take hours to produce in spec material because the plant was set up to expect a % of recycled materials. By shutting off the recycling flow during start up, the specification was achieved within less than half an hour.

Ideally quality standards should be the level the process capability (from assets that are 'fit for purpose') and this level should exceed what the customer expects. As TQM has taught us (and especially the 'world class' Japanese manufacturers) and what Six Sigma has rediscovered is that perfect quality can be free. Most businesses still believe that there is a trade-off or compromise involved with quality such that to get perfect quality then costs must go up.

This is a falsehood. Perfect quality products flowing through processes (without interruption) will generate more cash more quickly than less capable production systems (Mather, 1988; Schmenner, 2012). As will be discussed in Chapter 6, addressing minor product defects is the focus of Quality Maintenance such that, if the production process is in good shape, quality levels will be achieved more easily. Likewise, if defects begin to increase, it is an indication that the process is moving out of control and should prompt timely intervention to restore optimal conditions by engineering, quality and operations staff.

4.7 CHAPTER SUMMARY

Lean TPM provides a framework that is flexible to industry type and organisation structure. It has universal appeal as an improvement process and the Lean TPM approach therefore goes beyond the weaknesses of traditional quality improvements which have tended to lack the full involvement of the company team. The success of Lean TPM, and most other business initiatives, depends on the development of an agreed model of working and a change process to deliver it. To begin this process and model of change management it is important to lower the barriers between inwardly facing business departments (operations and engineering in particular) with outwardly facing departments such as the marketing function. Only by truly uniting these elements of 'world class' manufacturing will the necessary business ingredients and inputs be ready to exploit the opportunities that Lean TPM offers. The model also needs to be a 'living model' that is refined as each layer of waste is removed and incremental adjustments are made to gain stability, perfect the current product line-up and then look into what future manufacturing capabilities are needed to compete in the future. The Lean TPM road map helps to set out the models applied by successful organisations on their journey to world class performance. This leadership provides an important single change agenda across the value generating process. Without such a focus, individual agendas will confuse the change agenda.

Each management level and shop floor team has a different role as described by the Lean TPM treasure map. The key to success is establishment of a top-down and bottom-up partnership where everyone is committed to finding better working relationships. That is the key ingredient to raising standards and delivering outstanding performance for current and prospective customers to the business and its stakeholders.

In practical terms the top-down role can be defined as:
- setting priorities (consistency of purpose);
- setting standards and supporting delivery (collective discipline); and
- giving recognition (objective feedback).

On this last point, studies into teamwork indicate that the biggest failing teams attribute their failure to team leaders who let people get away with poor performance (Lafasto & Larson, 2002).

On the same basis the bottom-up role can be defined as:
- consistent application of best practice (capability)

- sharing of lessons learned (openness)
- problem ownership–continuous improvement (aligned goals)

The delivery of these roles is dependent on the joint development of:

- a clear compelling future model;
- a practical change process to deliver the new model;
- total immersion by management and shop floor to establish new, more productive working relationships.

The changes within the factory and redefinition of employee roles is a critical aspect of learning and moving from learning how to do 'things right, to doing "things better" and then on to the ultimate question of how to "do things differently'. At each stage of the learning journey, the role of management at each level will change and that is why it is important to plan the change and also to regulate it so that staff can adjust to the change. One of the aspects of the A3 change process which is often forgotten is the use of people and behavioural diagnoses (as well as tools like pareto) – these include driver diagrams and force field analyses – which show the problems with the existing management approach and behaviours and give an indication of what roles are needed to be implemented as a result of the change.

REFERENCES

Bicheno, J. (2000). *The lean toolbox* (2nd ed.). Buckingham: Picsie Books.

Brown, S. (1996). *Strategic manufacturing for competitive advantage*. London: Prentice Hall.

Cohen, L. (1995). *Quality function deployment: How to make it work for you*. Reading MA: Addison Wesley.

Cox, A. (1996). *Innovations in procurement management*. Boston Lincs: Earlsgate Press.

Christensen, C. (2002). *The innovators dilemma*. MIT technology Review June 2002.

Cummins and Townsend. (1999). Teams in agricultural education: an assessment of team process instruction. In: *Proceedings of the 26th annual national agricultural education research conference*.

Ficalora, J., & Cohen, L. (2006). *Quality function deployment and six sigma* (2nd ed.). London: Prentice Hall.

Jackson, I. (2006). *Hoshin Kanri for the lean enterprise: developing capabilities & managing profit*. New York: Productivity Press.

Kurogane, K. (1993). *Cross functional management: Principles and practical applications Tokyo APO*.

Lafasto, & Larson. (2002). *When teams work best*. Sage Publications.

Liker, J. (2004). *The Toyota way*. New York: McGraw Hill.

Mather, H. (1988). *Competitive manufacturing*. New York Prentice Hall.

Monden, Y. (1983). *Toyota production system*. Atlanta: Institute of Industrial Engineers.

Nakajima, S. (1988). *Introduction to total productive maintenance*, Cambridge, MA: Productivity Press.

Rich, N. (1999). *TPM: The lean approach*. Liverpool: Liverpool University Press.

Schmenner, R. W. (2012). *Swift even flow*. Cambridge: Cambridge University Press.

Senge, P. (1993). *The fifth discipline*. London: Century Business Press.

Spear, S. (2010). *The High velocity edge*. New York: McGraw Hill.

Standard, C., & Davis, D. (1999). *Running Today's Factory Cincinnati*. Hanser Gardner Publications.

Warren, & Westbook. (2009). Chapter 11: *A challenge to the critics*. Michigan law review (pp. 604–641).

Womack, J., & Jones, D. (1996). *Lean thinking*. New York: Simon and Schushter.

Chapter Five

Transforming the Business Model

5.1 TRANSFORMATION AND THE BUSINESS MODEL

Many of the 'silver bullets' offered and promoted to managers during the 1990s proved to be disappointing 'blanks' that delivered very little (Hill, 1985). A side effect was a belief that 'we have done this now'. So many businesses had done Lean and done Total Productive Maintenance (TPM)! Saying exactly those words showed just how ignorant some managers are. How can you do a journey of continuous improvement in one cycle? You cannot. The best businesses in the world remain humble when asked how much progress they have made – they would never suggest they had 'done it'.

Perhaps one of the reasons why Western companies thought they had 'done it' was the attention paid by original Lean businesses in making things easy – easy to do things right and easy to see when things are going wrong, easy to see the status of the production process with visual measures, easy to manage through simplified visual methods (standardised work and single-point lessons). But simple solutions do not in any way devalue the size of the problem or opportunity that needed to be conquered. Simplicity is an art – it is also a primary cause of sustainable improvement. Make things easy and change feels much less uncomfortable – make it easy and most people will understand it (positive behavioural economics!). The art of visual management, simple controls and escalating problems up through the management hierarchy are critical to simple and effective control of the business and its processes.

For some reason though, many Western firms simply could not sustain these good Lean TPM methods (and proven techniques). Their failure suggests that either these techniques were not really understood by the managers who 'bought them and did them' and were not implemented correctly or that the organisation did not have the necessary support structures that allowed these techniques to take root and grow (Womack & Jones, 1996). This chapter sets out the key change team roles to support the Lean TPM improvement process.

In most organisations, working practices have evolved over time, influenced by events which have long gone. A key challenge of the journey to 'world class' (via the milestone process) is designing a new model and deploying it, without risk, to the business (Hamel & Prahalad, 1994; Rother, 2010). Change is rarely comfortable so it is important to condition workers for (and reiterate the need for) the change and the opportunities it presents for the business and the individual – this form of empathy is at the heart of Lean TPM. Lean TPM reflects the principles of 'respect for the scientific way' but the best way is achieved through a 'respect for people' and 'leading with humility'. This may be a slower change process but it does ensure sustainability and it is customer focused and designed to deliver value.

5.2 LEAN TPM IMPLEMENTATION

A proven implementation programme is organised under four milestones as set out in Figure 5.1. The content of each milestone is discussed in the paragraphs below.

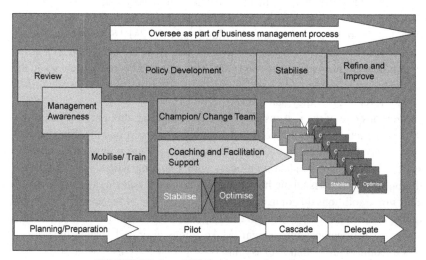

FIGURE 5.1 Lean TPM implementation milestones.

Planning/preparation

REVIEW

The aims of the review are to:

- understand current ways of working and assess strengths and weaknesses against TPM benchmarks;
- identify the overlap between TPM and current improvement toolkit to build on current good practices and address gaps;
- identify improvement potential; and
- meet key stakeholders to understand business drivers, future plans and priorities.

Table 5.1 sets out an outline review agenda. This usually requires between 2 and 4 days and includes discussions with key stakeholders to both explain what Lean TPM is and agree the scope of the pilot/learning to see programme.

TABLE 5.1 Lean TPM Review Agenda

Content	Notes
Business drivers, priorities and future goals	Understand current status, business drivers and future plans
Production review	Understand production process and ways of working
Engineering and maintenance review	Understand maintenance processes and ways of working
Financial data	Understand current costing and financial systems. Identify cost and throughput data for a representative period
Quality assurance	Understand quality measures and cost of quality
Analysis of available data	Calculation of overall equipment effectiveness and cost model development
Draft programme development	Set out proposed pilot programme
Programme review/refine	Lean TPM awareness and proposed plan review with key managers
Wash up	Plan next steps with programme sponsors

TPM, total productive maintenance.

Management Awareness

Following the review programme a management awareness/workshop session is carried out to:
- raise understanding of Lean TPM, its potential impact on the business and the implementation route map;
- refine the proposed Lean TPM implementation route map.
  Outputs from the review include:
- an assessment of the organisational performance against Lean TPM benchmarks;
- an preliminary evaluation of improvement potential in financial terms;
- increased understanding of Lean TPM and how it can be applied to meet the needs of the business;
- an implementation programme setting out roles, resources, timescales, awareness/training programme and an implementation quality plan.

Pilot programme

Figure 5.2 sets out a typical 13 week pilot programme. This includes work streams covering programme management, awareness and training and improvement team activities. Further details of each work stream is included at Table 5.2.

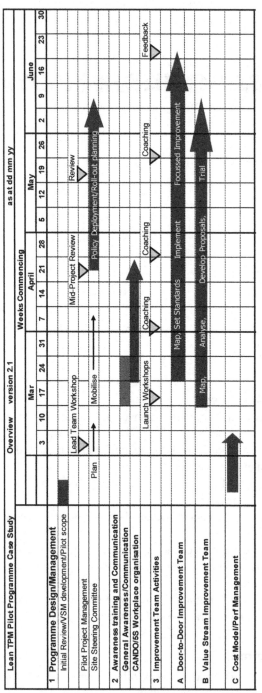

FIGURE 5.2 Lean TPM pilot timetable (case study example). TPM, total productive maintenance.

TABLE 5.2 Pilot Programme Work Streams (Case Study Example)

Project Stream	Notes
Programme management (including Lean TPM policy development)	Carried out by the lead team, this activity involves setting the direction for the project, monitoring progress, collating lessons learned and identifying and dealing with policy gaps and road blocks to progress. This also includes the planning of the roll-out process following feedback of the pilot programme. Key events include the mid-project review which is designed to ensure that the feedback session is a launch pad to the next stage of the programme rather than a decision point so that the programme momentum is not lost. Next potential steps could be to extend the pilot and roll out the programme to suspend/stop the programme.
Awareness and communication	This activity will be supported by 'train the trainer' coaching so that as much of the awareness/communication can be carried out by the internal facilitator and/or team leaders. The proposed awareness programme involves a series of 2-h training sessions for those not directly involved with the MacLean pilot. The content will include an explanation of Lean, the pilot programme and a practical exercise using CANDO/5S workplace organisation tools
Improvement team activities 3A and 3B	The proposed programme involves two types of improvement team as described below. The 'door to door team' will target improvements to the physical process improvement projects including the pasteuriser and filler optimisation. The 'value stream mapping team' will focus on the production systems improvement including the demand management, supplier evaluation, distribution optimisation and value stream mapping projects
3C Cost model	This is a key policy development area to support the tracking of business benefits and decision making relating to improvement options as well as confirmation of the benefits of value added from the process of the Lean TPM pilot.

A key milestone within this timetable is the mid project review approximately 6 weeks after the launch workshop. At this point the improvement teams should be in a position to explain what they have found and their implementation plans. This is also the point at which management attention should progress to defining roll out plans so that the feedback event can be a launch pad rather than a decision point.

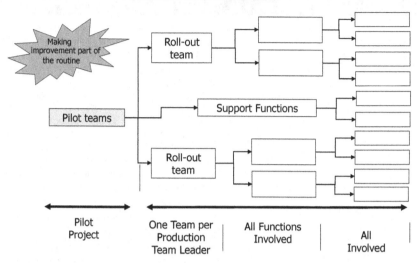

FIGURE 5.3 Lean TPM cascade (case study example). TPM, total productive maintenance.

Roll-out cascade

Following the pilot programme a formal roll-out cascade is used to deploy the lessons learned during the pilot in the form of policies, ways of working and accountabilities.

The 'roll-out' cascade covers the first milestone of the master plan and is completed when all personnel are involved. As shown in Figure 5.3, a typical roll-out cascade is designed in three phases. The phased approach helps to minimise resistance to change by engaging early adopters in the early stages. As each phase is completed, it also provides the sense of momentum to engage those who need more convincing. That is said, it is essential for operational managers including first-line managers (FLMs) to demonstrate their engagement from phase 1. For that reason the first phase of the roll-out cascade will concentrate on the direct shop floor activities to stabilise operations and provide a structured learning opportunity for FLMs. Those who cannot (with support) deliver improvement through their teams and achieve TRAC level 1A may be in the wrong job. The TRAC process is designed to reinforce proactive behaviours.

Confirmation of TRAC level 1A for each team leader is evidence based; quality assurance check that FLMs have learned what is expected from them. To achieve this basic level they will have had to have made and implemented plans to deliver improvement through a front-line team which is evidence that they have bought in to the process. If they haven't, don't fudge the results, find out why and deal with it. Encourage and help FLMs but also hold them accountable. At the same time hold their managers accountable for giving 'first line managers' support. Make sure they remove barriers to progress. The first phases of the roll-out will tease out how to get the organisation to accept improvement as a part of everyone's job. Full immersion in this goal is essential to establish the patterns of behaviour that becomes the elevator to industry leading performance.

Once the first phase of the roll-out cascade is confirmed, extend the cascade to include support functions (the 'indirects' as people call them because it is here that much of the business delays and mistakes occur). It is important to 'vision' what these processes will look like in the future – to look forward to give the business time to adjust to the future competitive capabilities needed to deliver customer value profitably.

For example, a capital project or technical improvement to be delivered in the future will need to be introduced in a way that anticipates and facilitates a new collaborative and team-based approach (involving many departmental stakeholders and the operational teams). This requires the involvement of personnel/training and project engineering as key project leaders. The project might also support changes in the value stream that requires the involvement of the planning department and customer services function (stakeholders). In other words, the impact of some decision taken today will not be felt until later and a simple framework is needed to guide the way. It is important therefore that mistakes, at the early stages, of such a process are minimised to avoid costly confusion, rework and chaos towards the end of the project.

The future-state vision is therefore a seamless integration of employees using formalised processes to achieve customer-valued outputs. The ability to set out a 'future-state' requirement for the business and to develop the necessary support activities and education is paramount for 'world-beating' businesses. This is not that surprising really as these processes are keys to ensure that everyone knows where the business is going and how it intends to get there. For traditional businesses, hardship and manufacturing problems are associated with budget cuts and not, as high-performance businesses practice, greater and greater levels of training towards the final goal. There are no compromises in delivering the future state, and central to this 'delivery' is getting all levels of personnel to learn, experiment as teams and reflect upon what has been achieved and the extent that the gap between today and the future state has been closed (Dimancescu, 1992).

5.3 LEAN TPM IMPLEMENTATION ROLES

Throughout this book we have used the term 'value' as something we generate for customers. We have used it to describe processes that convert materials into saleable products, to describe processes that compress the time between receiving an order and fulfilling it – defining value from the perspective of the customer (Lean principle number one!). There is, however, another dimension to 'value' that is rarely discussed in the modern management literature and this is the value of organisational roles within the firm. An understanding of these 'value roles' is an important context that has a relationship with successful and sustainable process improvement.

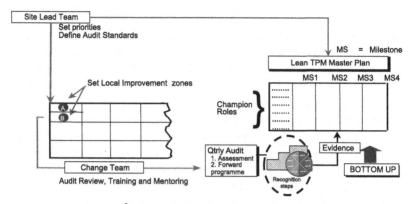

FIGURE 5.4 Assessing top-down leadership performance.

The two generic work flows which add value as part of the top-down and bottom-up Lean TPM process are owned by the site lead team (top down) and the change team (bottom up). Collating the bottom-up audit results is used to assess top-down leadership progress through the Lean TPM master plan milestones as set in Figure 5.4.

Top-down site lead team

1. **Senior Managers**
   a. identify winning business and functional strategies;
   b. support this with a coherent set of priorities and standards;
   c. consistently manage results and recognition systems;
   d. align accountabilities and build the capability to deliver them.

   This includes key areas of policy such as focussed improvement, safety, environment and administration.

2. **Heads of Operations** whose role spans the planning organisation and control of the transformation process and deliver quality, cost and delivery performance. These employees determine the rate of change and they control the systems that deliver customer value.

3. **Heads of Maintenance** whose role spans the planning organisation and control of asset care and optimisation of current technology with the additional responsibility to eliminate the current issues of the current technology when designing or procuring the next generation of technology. The role includes quality maintenance and demands that the head of maintenance works with the human resources department to determine the 'future maintainer role' and future contracts of employment (training needs analysis and designing in value/continuous improvement into the technical job role).

4. **Heads of Skill Development/Training Mangers** who have responsibilities that cover the provision and operation of systems and processes to raise capability across all levels in the business. The training managers work on the current skill set and on turning training into a profitable activity for

the business. Training includes mandatory and compulsory training (and monitoring) as well as equipping staff with skills that increase the person's contribution to the inter-personal change management processes of the business.

5. **Heads of Technical Management/R&D** whose role covers the provision of technical excellence and flawless delivery of products, equipment and processes through early equipment management and the compression of time between equipment design and the point at which the equipment is at the design speed/quality and earning its living for the business.

Bottom-up change team

This includes local policy champions:

6. **First-Line Managers** who lead the shop floor teams, establishing local policy, developing the potential of the team and it's team members and supporting the continuous improvement process. These managers are critical and they maintain the discipline of the teams on a daily basis. It is important that the daily management activities of these staff are routinized to allow the person to get involved in improvement activities;

7. **Multiskilled Shop Floor Teams** whose role is to make good production and to engage in problem-solving where issues are detected with the task or production process. These staff are important and their skills need to be carefully developed in a sequence of learning that heightens their sensitivity to the production system and in particular its abnormalities and signs of deviation.

Support Team

Finally there are two key supporting roles including:

8. **Continuous Improvement Manager/Facilitators** to support the planning for, implementation and conditioning of, change. Although this role is critical to a successful improvement process, the role has no routine activities. This is a role concerning with building new ways of working, improving collaboration, stabilizing the new approach and passing it on. In today's business world of constant change and increasing challenge, the best companies make this a full-time role.

9. **Specialists/Key Contacts and Support Functions** that support the value adding process. Their role is to support the capture and transfer of technical knowledge, target/remove bureaucracy and raise capabilities and create cross-functional collaboration without boundaries. These individuals include health and safety staff, calibration specialists, technical analysts and other specialist roles.

These are the basic building block roles of the change team. The enormity of change is too big for an individual to control for most businesses and therefore a collective and team-based approach is needed. In the early stages of developing your own improvement process, such an approach (and the amount of details involved) can seem confusing and unhelpful. But in reality, there is a logic that underpins a master planning process based on the experience of successful change programmes.

5.4 PROGRAMME MANAGEMENT

Although the Lean TPM programme steps are well structured, it is important to integrate programme management as a core business process. This helps to sustain goal clarity for the business management teams and shop floor workers alike. The following decision steps provide a process to establish top-down roles to support the bottom-up delivery.

Step 1. As part of the 3- to 5-year look-ahead within the annual business planning process, define the results required to pass through the next two master plan milestones over the next 3–5 years. Define the milestone exit criteria in outcome terms. Set out realistic, achievable but stretching goals for the next 12 months based on an assessment of the Lean TPM treasure map priorities.

For example:

- Deliver zero breakdowns.
- Halve lead times and inventory.
- Minor stops to 1/10 of previous levels.
- Improve quality consistency.
- Reduce new product introduction times.
- Improve morale.

Step 2. As part of the annual business plan, break this down into steps of around 3 months duration and define the competencies, development and practical projects required to deliver that goal.

Step 3. Allocate each project to a senior management team member.

Step 4. Mobilise the projects as part of the policy deployment process

As all of the management team will have an influence on issues of pace, priority and resources allocated, each member of the management team should be engaged with this programme and also aim to establish a single integrated improvement process.

Setting and raising standards

A key outcome of the policy deployment process is the definition of guidelines or standards to be implemented as local policy by first-line management for including:

- the analysis of improvement potential, setting of priorities and communication of goals. This includes the use of visual management and two-way information exchange to engage all personnel under a compelling vision and help them to make sense of the world from their perspective;
- best practice operation definition (e.g. start up, steady state and close down) including cross-shift standardisation;
- development and application of lifetime maintenance processes to establish basic conditions;

- skill development and delivery processes;
- development of a technical trouble map, knowledge base and objective decision-making processes to deliver products, equipment and processes so that they achieve flawless operation from day 1.

This deployment of local policy is also the mechanism for transfer of best practices between departments.

- **Policy standard** (set by management): Welding equipment will have a routine of daily cleaning and inspection checks.
- **Local standard** (set by shop floor teams):
  - change CO_2 wire,
  - check pressure setting,
  - clean table and tooling,
  - check clamp head security,
  - check for air and water leaks,
  - check torch and harness security, etc.

5.5 CHANGE TEAM

The collective FLM role is the deployment of business-wide policies in their local area. This involves some sharing out of geographical improvement zones to create areas of physical responsibility for teams and managers (it may even go as far as colour-coding areas of the factory to denote the team responsible for everything in that area). It will definitely involve enforcing workplace disciplines and reporting non-compliances. The development of detailed local policy can then be shared across teams and across shifts to reinforce ownership and responsibility of teams and key workers. In addition, the FLM needs to support the evolution of the team towards a desired level of self-management by deploying routines to teams based on their competence and abilities. During this process, the FLM role changes from a role of dictating and directing to coaching, supporting and finally 'hands off' empowerment (when teams have proven they can adhere to the agreed scientific procedure – see Rother (2010)). The stepwise development process set out in the table below guides this task and provides the foundation for self-managed teamwork such that teams can:

- identify potential improvement opportunities,
- secure 'zero breakdowns' through effective asset care, and
- optimize process capability through the application of problem prevention.

Simultaneously, coaching by senior management supported by facilitation provides the foundation for FLM leadership development through the stages of:

- directing the change process,
- coaching in the new behaviours and responsible empowerment,
- supporting development of team self-correction capabilities,
- delegating responsibility, and
- Auditing and constructive criticism.

Although often this is supported by formal leadership training and/or one-to-one coaching, the need for managers to coach their direct reports cannot

TABLE 5.3 Bottom-Up Audit Coaching Level Overview

Level	Guiding Concept
Teamwork basics A (milestone 1)	Basic information recording and communication, routine problem definition/frequency, initial cleaning of workplace/equipment and formalization of critical procedures
Teamwork basics B (milestone 1)	Standardization of basic operational and maintenance practices across departments and shifts
High-performance teamwork A (milestone 2)	Simplify/refine practices to reduce human error/unplanned intervention and release resource/energy for improvement activities
High-performance teamwork B (milestone 2)	Raise awareness of inspection needs to provide early problem detection capabilities. Deliver zero breakdowns and stable operation
Cross-functional teamwork A (milestone 3)	Build the technical capability of core/workplace personnel and reorganize to delegate routine activities to them. Develop technical team capabilities to focus on optimisation/stretch targets
Cross-functional teamwork B (milestone 3)	Bed in new ways of working and identify how to deliver optimum running, improved quality consistency and reduced variability
Cross-functional teamwork A (milestone 4)	Deliver and maintain optimum conditions
Cross-functional teamwork B (milestone 4)	Define and strive for next generation of zero targets (e.g. defects, inventory)

be side stepped. Total immersion is necessary to establish new proactive working relationships based on high communication and trust. In parallel, Lean TPM provides a stepwise bottom-up team-based learning/development process for shop floor teams. Combined with the top-down coaching by senior managers and first-line management leadership role, this provides the mechanism to deliver progressively increasing levels of team empowerment.

Table 5.3 below sets out this stepwise team development and learning framework in terms of the capabilities that are mastered at each stage. In Lean TPM speak, below are the steps of autonomous maintenance. They have been explained in more general terms in Table 4.1 because of their value as a high-performance teamwork development process.

5.6 OPERATIONS TEAM

The 'master plan' and bottom-up learning process for shop floor teams set out what can be achieved and what is desired from each team member within the factory. This reinforces the discipline of the following:

- Routine operation to make good products and ensure customer satisfaction. This is the primary role of the operations team and its members.

- Improvement/learning from experience.
- Feedback and monitoring so that problems can be escalated by 'managing upwards' and calling for help to eliminate the barriers to 'zero loss' performance.
- Engaging in 'self management' where trained and safe to do so.

These characteristics are the bare minimum of the operational team member's role. The art of Lean TPM is to increasingly blur the edges of traditional job descriptions such that what is regarded as 'routine' by functional support areas to the operations teams can be packaged and deployed to the team themselves. The movement of quality-assured processes to the team level therefore heightens the sensitivity of the team to abnormalities and increases both skill variety and the empowerment of the teams to act as 'businesses within a business'.

To fully own a process, it is important to make things simple! That is, to make a process 'easy to do right and difficult to do wrong'. The art of simplicity must be applied to the team level and used to allow the team to manage easily without unnecessary burden. Simplicity allows checks to happen more often and simplicity releases valuable time for staff to concentrate on other things – such as improvement. A great example of simplicity is the use of visual indicators on the locking nuts located on the wheels of heavy goods vehicles (see Figure 5.5). These indicators are bright green or yellow plastic and cost a fraction of a penny but they perform a critical role – they are a safety device. On a wheel without these indicators it is difficult to detect when a nut is coming undone until it falls off. If the nut falls off then potentially a wheel and an entire vehicle and driver may be damaged and lost. These devices are placed over the nuts of every

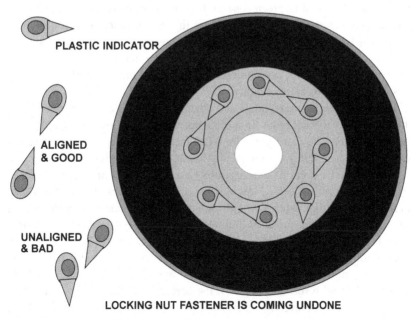

FIGURE 5.5 Simple but effective – the process calls to you.

bolt stud for every wheel. The pointed end of the indicator should point at the pointed end of an adjacent stud – in this way if the nut was to work loose then the indicator would show it. In the figure below, all indicators are flawlessly alligned except two. As the driver enters and descends from the vehicle, he or she can see instantly that something is wrong and it is likely that the bottom right indicator is showing the nut is working loose. The driver can therefore tighten the nut before it works its way off and a potential catastrophe happens. Thankfully visual indicators and visual measures can be used to make factory life easier. Easier to do right with minimal interruption to the working day.

The typical tyre is checked many times during the day – up to around 20 times a shift – as drivers get in and out of their vehicles. Now operating a machine in a factory has much less time when the operator is away from the machine (or in this example inside the driver is in the cabin of the machine). Instead the operator is walking around the area and passing equipment every couple of minutes or so. As such it is possible to make checks very easy without interrupting the value adding cycles of the operator. Checking the dials on the machine that monitor air pressures, lubrication oil levels, water flow rates and other such activities is an activity that must happen on a daily or more frequent basis. It is possible to make these indicators very user friendly and to use visual indicators to show abnormality and then to develop processes that either 'stop the line' instantly on detection or allow the issue to be escalated to the competent person in the organisation who can solve the issue. For instance, a pressure dial is a critical item – too little pressure and the machine will not operate correctly and produce good products, while too much and the machine may catastrophically fail and even cause injury. Simple colour-coded indicators with red markings to show unacceptable levels and a green zone to show the normal operating range (when the machine is in use) make it easy to see that the machine is in control – or escalate the issue if the machine enters the red zones. The same logic is applied to the revolution counter on a car with excessive revolution range being shown to the driver in red or the petrol/diesel tank level and the red minimal warning lines on the visual dashboard indicator.

If the operator has many dials and other indicators to check then it is possible to make every dial point (in normal operating conditions) directly up – as if pointing to the 12 o'clock position. When faced with many dials it is therefore very easy to detect if any individual dial has deviated (see Figure 5.6).

Following the logic of making jobs easy to do right and difficult to do wrong is the development of routines for operator checks. These checks are similar to those that a pilot would undertake before taking a multimillion dollar plane up in the air with hundreds of passengers on board. A simple example is the use of five-by-five checks (see Figure 5.7). The five-by-five check is quite simple to understand – five checks a day for 5 days! Each check is a single piece of A4 paper that is designed to show the checks to be undertaken with lots of photos of the actual check to show the standard and the means of conducting the check or doing the activity safely. These

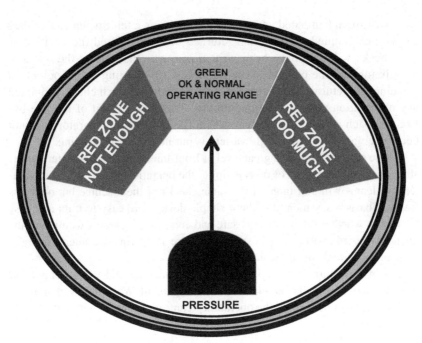

FIGURE 5.6 A pressure dial.

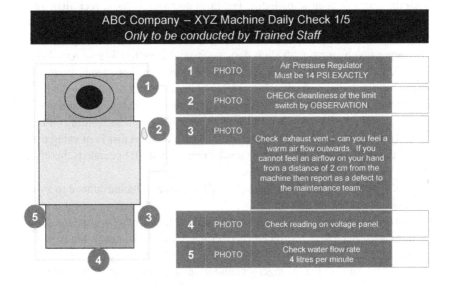

FIGURE 5.7 A daily check.

sheets are often laminated and are completed using a felt tip pen so that they can be photocopied, cleaned and returned to be conducted the next week. The checks will also have been written in a way that allows the operator to walk through the process in the simplest way – starting at one point of a machine and returning to that point having conducted each check and noted any observation that must be undertaken. The other benefit of the five-by-five approach is that the cards can be shuffled so that the operator does not become complacent and a little variety is put into the daily routine.

These systems provide a great level of front-line control and transfer ownership (and good standards of ownership) to the operations teams. Central maintenance teams will still monitor the same checks of the machine but on a less frequent basis – say monthly. These simple devices are easy to train and easy to witness whether the operator is safe and competent in their execution to the quality standard expected. These documents are also living documents and will be improved over time to include more checks.

Again we return to the softer aspect of managing and the development/integration of new skills via relationship management. We will now explore the major relationships held by the operations teams and support functions.

5.7 SPECIALISTS

Team leaders and team members do not exist in a vacuum and instead have a network of support specialists and key contacts upon whom they call when they reach the limits of their knowledge. For most businesses, these **technical specialists** will be internal coworkers in different departments but it is also increasingly true that asset manufacturers and other suppliers are also taking a greater and deeper involvement with shop floor teams and their training (external specialists). Internal specialist technical staffs hold specific knowledge bases, for which they have attended specialist training/college courses to become qualified in what they do. This is an interesting point to ponder – a maintainer cannot only operate a machine but has received technical knowledge, understanding on how the machine functions, etc. The value of this form of specialist is in using their trained diagnostic skills to help them teach operations staff. Specialists therefore provide assistance in the form of:

* establishing a 'trouble map' of where efforts need to be introduced to assist material flow performance;
* horizontal empowerment (standardization) such that other teams and functions are trained to the correct level of efficiency to do the new roles expected of them;
* providing knowledge management and training to high standards (including the auditing of standards to ensure they are efficient and safe);
* the 'flawless' implementation of change including designing improvements on behalf of the team to ensure high levels of material flow.

The commercial function and **commercial specialists** also provide the operations teams with a range of benefits and outcomes needed to improve and support the self-management/regulation of team performance. It may be strange to think that this relationship exists as many factory teams have little to no understanding about the customer but do understand the processing requirements needed. It seems illogical that the 'customer focus' has not been shared to this level. As such, this operations–commercial relationship offers many benefits including:

- financial awareness so that the teams can make decisions that improve the commercial performance of their area of responsibility;
- life cycle cost focus to ensure a long-term pay back to the company that can be used to ensure the 'correct' rather than 'reaction' decisions are made to the production process;
- cost deployment to focus the improvement efforts of the teams and engineers;
- customer awareness and 'voice of the customer' reviews such that changes in the needs of the customer are communicated to the various teams to assist their approach to improvement activities;
- monitoring the market for innovations that could be employed by the firm including the availability of new technology processes and 'best practices'; and
- ensuring that the business is fully compliant with the safety, legal and contractual requirements of manufacturing.

5.8 FACILITATION

Having access to technical skills and the willingness of the support functions to assist the development/empowerment process is just one aspect of migrating skills to the person at the point of value adding. Traditionally, specialists speak with a technical vocabulary and they find it hard (not to say frustrating) when attempting to describe processes and subjects to non-technical people especially operations teams. Without a common understanding, willingness to help will soon evaporate if the technical person is not confident that the operations teams understood and whether can apply the new knowledge (no matter how willing they are to lose a routine, mundane and nonvalue added part of the technician's job). As such, facilitators are needed, and even the smallest of businesses will need a facilitator, to provide the control of the master change plan and ensure that technical and inter-personal training is undertaken to take the teams to the next stage once they have mastered the current stage. Facilitators are therefore instrumental to the management of change velocity and provide the following roles:

- Management of the master plan for each area of the factory.
- Continuous improvement and technical competence needed to train the various factory teams.
- Programme design and co-ordination of change throughout the firm including indirect departments.
- Review the number of suggestions developed by the shop floor teams and also to monitor the implementation stages of these suggestions.
- Highlight road blocks and support resolution of these issues.

- To maintain the catalogue of improvement activities and maintain all documented procedures (including ensuring these are safe, efficient and are recorded within the quality management system of the firm).
- Maintain a record of training and competence levels achieved by each employee during the change process.
- Provide summary reports for management in terms of progress and improvements achieved,
- Control of training budgets and budgets for external specialist support needed.

Facilitators are also heavily engaged in the promotion of change and the motivation of the teams in the factory as a precursor to sustainable change and, as the name suggests, do not necessarily have to be the experts but must ensure they have a working knowledge of the technique in order to energise the team members and facilitate the change process appropriately. One of the important roles conducted by the facilitator/facilitation team is their responsibility as performance monitors of the change process. As such these individuals are instrumental as a linkage between the management and the overall performance of the new business model in terms of progress. In this manner the facilitation team and factory management are likely to engage in the development of key business reports which will assist the annual rounds of business strategy development and execution. These reports will typically include:

- the development of a 5-year master plan,
- the quantification of business losses and potential,
- the justification and prioritization of change efforts at the team level,
- an annual analysis of costs including the targeting of costs to be reduced and some influence in the budget process,
- the preparation of quarterly feedback reports which detail the progress of the teams in meeting the business strategy, and
- the development of case study learning materials drawn from the teams and used for illustrations for management of the profitability of the change programme (and consequently the benefits of an aligned marketing and operations strategy).

5.9 CHAPTER SUMMARY

The definition of job roles and the assignment of activities to these personnel (to support the Lean TPM change mandate) are critically important to secure a change in the business model. For most businesses some of this infrastructure will exist but has not been 'joined up' in a way that supports 'top-down' strategy deployment and a 'bottom-up' active participation in these changes. A change programme that is not structured effectively will certainly reduce its chances of success in the same manner that a lack of sufficient resources will be a major inhibitor to effective change. Above all, the appointment and training of facilitators is a visible sign that the management of the firm is serious about change and supporting the new business model itself. In short, these features ensure that change is successful and avoids the most basic forms of why programmes fail (lack of time, lack of serious management intent and lack of support).

The analysis of roles and responsibilities is also important in identifying, for each group of personnel, the value they will be expected to play in the future as part of the growth strategy. A perfect strategy will not be realised if the skills throughout the firm are not aligned and focused upon improvements that eventually contribute to the efficiency and effectiveness of the firm and this demands a rethink of roles. However, as this chapter has portrayed the roles exist within the firm and do not tend to involve a restructuring exercise (with the associated delay and time lost as individuals adjust and make sense of their new roles as a result of traditional restructuring). Instead the master plan is a refocusing of what adds value to the primary concern of keeping materials moving, increasing value added and collapsing the payment cycle. These roles exist but, as typical of most traditional management practices, have neither received much attention nor alignment of roles to form a cohesive and robust change management structure.

The alignment of these roles and appropriate skill development, of facilitators and the facilitated, with the future-state design of the firm and the development of regular periodic reviews at the close of each Lean TPM milestone ensures that skill provision and the incremental mastery of each Lean TPM pillar maintain an ongoing process of role transition and empowerment of operations personnel (and the empowerment of indirect staff released from these low-value added routines).

Central to the development of meaningful relationships that support high performance and sustainable improvement is the correct selection of key measures. Measures help reinforce dependency between staff – that is to say reducing the costs of maintenance cannot be conducted entirely by the maintenance department and its staff – it requires routines of control to be adopted by operations teams. Visual key performance indicators that correctly identify the need to manage quality and the timely delivery of activities are critical to keeping costs down but more importantly to keep staff talking about the business and exploring new ways of improving. When you realise that your performance is inextricably linked to the performance of others (including suppliers) then you realise that investing in relationships and collaborating is the only viable way of getting to world class levels of performance (Greif, 1991; Ortiz and Park, 2010).

REFERENCES

Dimancescu, D. (1992). *The seamless enterprise*. New York: Harper Collins.

Greif, M. (1991). *The visual factory*. Portland: Productivity Press.

Hamel, G., & Prahalad, C. (1994). *Competing for the future*. Boston, MA: Harvard Business School Press.

Hill, T. (1985). *Manufacturing strategy*. Basingstoke: MacMillan.

Ortiz, C., & Park, M. (2010). *Visual controls: Applying visual management to the factory*. New York: Productivity Press.

Rother, M. (2010). *The Toyota Kata*. New York: McGraw Hill.

Womack, J., & Jones, D. (1996). *Lean thinking*. New York: Simon & Schushter.

Chapter Six

Process Stabilisation

6.1 STABILISING PROCESSES

At the heart of Lean total productive maintenance (TPM) is the optimisation of the value stream (Womack & Jones, 1996). As set out earlier, striving towards the Lean principle of perfection is a systematic journey of learning and developing the organisational capability to:

1. Establish the company-wide best practice recipe for low-inventory, high-flow stable operation (Policy deployment of Lean TPM).
2. Lock in the 'best practice recipe' to deliver 'zero breakdowns' and self-managed teamwork.
3. Identify the route map to release the full potential of the current operation and build the foundation to match and exceed future customer expectations.
4. Change the competitive landscape and lead the customer agenda for products and services (build current and new product competitive offerings).

6.2 ASSESSING THE GAP

The first side of this organisational 'Rubik cube' to solve the concerns associated with closing the gap between current state asset performance and determine the reliability envelope or what it could consistently achieve (by design). Put another way, the difference between a machine (or process operating) without breakdowns and the current state. Getting in control of technology is the first bottom-up target. This provides a 'hands on' learning programme for shopfloor teams and support functions. For managers, this learning is captured by the use of 'value stream maps' that show the commercial impact of asset reliability. Understanding these issues raises understanding of the causes of failure, how to stabilise and extend component life, detect signs of failure early and take prompt timely action. This vastly reduces the amount of unplanned interventions and compresses the time needed to deal with abnormalities as they occur. As each solution is developed, it reinforces learning and the drive for future improvement activities because teams can see the benefits of their problem-solving actions (each solution and learning process should be captured using a method known as the A3 process which we will explore later in this chapter).

Activities to raise the understanding of customer value, establish current benchmarks for levels of loss/waste and prioritise improvement activities are carried out in parallel and as part of the policy deployment process used to set the gap and stretch objective. Awareness of the gap provides the learning challenge at the management level (the only level that can implement system changes to eliminate organisational and inter-organisational wastes). This often overlooked element of the improvement process provides:

- identification of nonvalue adding activities and implementation of low-cost or no cost improvements, and
- the foundation for the top down/bottom-up partnership as discussed earlier in this book.

The former activity should improve added value per labour hour by 10–15%. This increases the importance of the latter because if these milestones are going to generate a 10–15% increase in capacity with 10–15% less labour interventions then the question is how will management turn this into commercial gain?

Growth strategies produce around five times the return of ones based on downsizing/head-count reduction. In addition, they have a positive impact on staff motivation. The 'growth' solution will be different for every organisation but without doubt using the Lean TPM process to underpin a customer-driven business strategy is the most powerful way to assure success. As such, Lean TPM is most likely to be one of the competitive capabilities and improvement activities that will be undertaken by a manufacturing business and will certainly be identified by any business that engages with the policy deployment process.

6.3 UNDERSTANDING THE VOC

'World class' organisations are outwardly focused and understand the customers they serve and hope to serve (Womack & Jones, 1996). The best way to conduct this analysis is to bring together all the functions and departments that must combine to deliver a typical order from the point of customer interaction to the final despatch of the product (a door-to-door approach). Bringing all managers together is important – if only for awareness raising – it is a fundamental approach within total quality management as we know and it helps managers to learn about their systems and why some departments are so passionate about what they are doing. Most notably this approach unites sales and supply chain managers with the Lean TPM programme. Such an approach requires the combination of representatives from the sales functions, the operations, quality assurance, maintenance, production planning and supplier scheduling at the very minimum. Each of these managers has, to a greater or lesser extent, an influence upon the performance of the firm and the customer service delivered to its customers. These business functions need to be brought together to act as a single decision-making team that designs the future capability of the firm. It is also this team, through a process of mutual dependency that must understand both the external needs from the market and the internal demands and constraints facing all other departmental managers. It will not be a surprise that these managers control the resources needed to meet the four key milestones of Lean TPM, and these managers must stay together throughout the Lean TPM learning journey. No modern world class company has ever gained their capabilities with a fragmented organisation that defends insularity and departmental boundaries – quite the opposite.

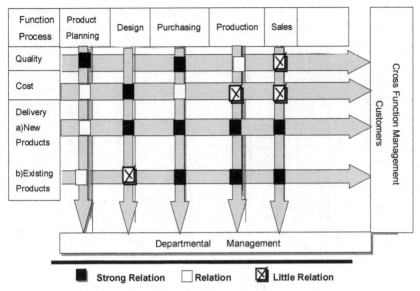

Function Process	Product Planning	Design	Purchasing	Production	Sales		
Quality							
Cost							
Delivery a)New Products							
b)Existing Products							

Departmental   Management

■ **Strong Relation**   □ Relation   ☒ **Little Relation**

FIGURE 6.1 Cross-functional team process impact.

World class organisations have also learnt that the 'command and control' structures associated with the mass production era are no longer sufficient to increase the decision-making capability and responsiveness of the firm. As such, world class organisations design systems and do not change any part of the production system without understanding the impacts of these changes on all other parts of the firm. Figure 6.1 sets out the cross functional impact on the key performance areas of Quality Cost and Delivery. This also highlights the need for a cross functional team approach to deliver lasting performance improvement.

With this team in place, finding out the major customers to the business is relatively straightforward. Gaining an insight into the most important customers can be achieved from a simple pareto analysis of annual volumes bought by all the customers of the firm. To understand these main process flows and their associated products is important in order to gain control of the products that represent the 'runners' (high volume) and raise the most income for the business. An example of demand profile analysis is shown at Figure 6.2. Here products to customer A account for 55% of total demand. There is little point in beginning a process of stabilisation by starting with low-volume products (this can come later) and initially the focus of the process will require attention to the main products, customers and support processes needed to produce the mainstream stock keeping units. These products also recover, due to their volume, the most of factory overheads.

Having identified the customers to the firm, and before any internal analyses are conducted it is important to determine the buying criteria of these businesses. For many firms this is the first time that managers, from many internal departments, have gathered together to understand the true 'voice of the customer (VOC)'. Instead in industry it is often the case that only the marketing and sales functions collect customer information and tend not to share it with those managers charged with converting materials into products that

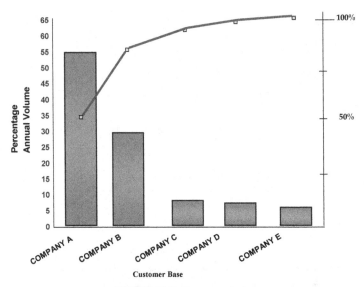

FIGURE 6.2 Pareto analysis.

customers want and are willing to pay for. The problem with both the TPM and lean approaches is all-too-often, and despite an implicit acknowledgement that satisfied customers result from the actions of many business departments, the majority of manufacturing businesses still do not share this type of information. As such it is no wonder that process optimisation rarely occurs and the first 'rule' of lean thinking is treated as a marketing not a business priority.

Finding the 'VOC' and translating it into a value proposition, demands questioning and seeking out the things customers want and clearly prioritising these needs in order to build a production system that is capable of consistently achieving these needs. For all customers to manufacturing firms these needs can be summarised into a range of features, many of which go well beyond the mere notion of 'price'. An example of a VOC analysis is shown at Figure 6.3. The first column shows the importance of the feature to the customer. The right hand column compares current performance with competitor offerings. It shows that currently important customer expectations regarding lead time and availability requirements are not being met. The challenge for this company is clear and so are the priorities for the cross functional improvement teams. To begin the VOC process it is worth thinking about two aspects of 'value', identified by Hill (1985), as:

1. What features are needed to 'qualify' our business to engage in trade with the customer base? These features are the basic implicit and performance features needed simply to meet current customer expectations of good service.
2. What order winning features will differentiae our offering from that of our competitors. These features could include providing customer order qualification features at much better levels than the rest of the competition, innovations or extras in the product-service bundle.

For the purpose of process stabilisation both order qualifiers (the most basic of expressed customer needs) and order winners represent a solid foundation

Customer Want	Score/10	Rating Against Competitors
		Bad OK Good
Improve quality of product/service	8/10	
Make dependable delivery promises	5/10	
Offer durable and reliable products	10/10	
Offer fast deliveries	6/10	
Make rapid volume changes to meet demand	2/10	
Lower lead times	10/10	
Offer customisation	7/10	
Offer rapid design changes	9/10	
Offer full product availability	10/10	
Introduce new products quickly etc.	2/10	

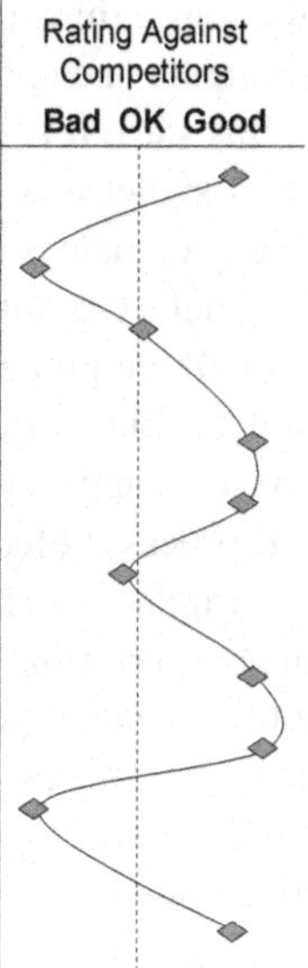

FIGURE 6.3 Competitive analysis.

upon which to build a high-performance value stream that gives customers their needs at a profit to the firm. Most customers value basic outputs in the form of a stated level of quality performance, delivery performance and price. These can all be plotted and as all manufacturers know, it is possible to create production systems that can reconcile these demands. For instance, to build a production system that offers the highest level of quality will reduce costs (due to less failures, better due-date performance will result and productivity will increase). However it is also the case that new 'buying criteria', such as environmental performance, may also be stated (often an order winner for customers keen to show their concern for the environment). At the heart of these debates will be quality, delivery and cost (QDC) as the basis of customer service. In modern times, it is delivery performance that has yet to be fully mastered and optimised to allow good-quality material flow to be combined with an ability to reduce costs and lead times. This round of preliminary analysis is very useful and will generate a series of performance indicators that allow the firm to benchmark with its chosen sector (as well as identify a number of existing and new customers who would be attracted by performing at this level).

However, understanding the current 'VOC' is not enough and, given the length of time needed to change a production system it is necessary to debate and discuss the trends and future-predicted performance requirements needed by the existing customer base. This round of discussions merely identifies what performance is needed to continue to qualify for work with existing customers and therefore should also include discussions of what the optimal level of customer service would be. A note of caution must be sounded here. It is not desirable to think of these order qualification and winning criteria over a greater period than the next 3–5 years because any greater amount of time is subject to vagary and many imponderables (including things like regulations, new technology, etc.). Thinking over too long a time frame is unlikely to yield meaningful information with which to reconfigure and optimise the internal processes of the firm.

At this stage it is important not to get bogged down in debating the minutiae of future-predicted customer needs – trends are important though. Consider broad improvement themes to support the development of in-house capability such as:

1. Zero defect quality products leaving the factory with zero losses to quality through the process. Whilst many manufacturers will have achieved a zero defect rating with their customers, few have achieved it at the internal process level and instead rely upon inspection, by officials or operators, to 'filter out' process defects. Even companies at 'parts per million' (PPM) levels of performance often find that the internal production process has not mastered quality performance to the PPM level.
2. 100% due-date performance in a lead time that is half of today's level or stated customer expectation as a minimum.
3. A reduction of production costs and prices charged in the region of 3% per annum.
4. A greater range of products and a much shorter time to market (from the drawing board to available to order).

6.4 VISUALISING THE VALUE STREAM

The next stage of Lean TPM is to visualise the value stream through which the value must be generated without waste, delay or interruption. This again is an exercise well worth conducted with the entire representative of middle managers whose functions influence the QDC performance of the firm.

The objective of this stage is to draw the basic blue print of the manufacturing process (and its associated information flows) as a series of connected diagrams. An example of a value stream map is shown at Figure 6.4. This is a current state map designed to set out the relationships between information and physical flows. Having conducted these analyses and 'maps', the next stage is to identify the problems with the existing system and create a series of potential future state manufacturing models. The combination of these current and future state value stream maps provides a benchmark of what currently happens and what the optimised manufacturing process value stream could be like. The latter is obviously the focus of the continuous improvement activities of the firm and is based upon the performance measures that have been predicted in terms of basic qualification criteria and the order winning level needed to maintain a 'manufacturing-led competitive advantage' over other producers in the same sector. The process of mapping the value stream involves:

1. Identifying the main manufactured products by the factory in terms of a pareto rating and importantly the main products bought by the major customers to the company that were discussed in the previous stage. This analysis provides an understanding of A Grade products (those products with the highest annual sales and regularity of production), B Grade products (medium sales and production regularity) and C Grade products (the tail of the product offering with very low volumes and infrequent demand).

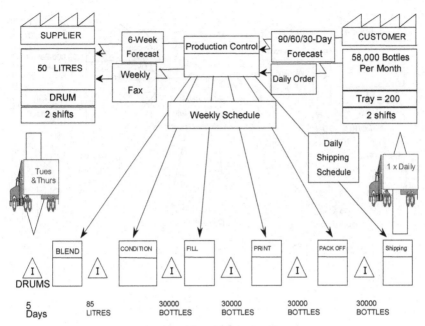

FIGURE 6.4 Current state map.

A general 'rule of thumb' is that the A Grade products will account for 20% of all products but 80% of annual volume whereas the C Grade products will account for a lot of the products but only a very small percentage of annual sales). This analysis is critical (even though it is typical for all these types of products to be subject to the same QDC). The purpose is therefore to design a production system that allows the 'A Grade' products to be manufactured with zero losses and let the other products benefit from this process design.

2. To list the sequence of manufacturing stages through which the product is made and converted into a finished good and to identify, for every stage:
   a. the available minutes in a working day,
   b. the cycle time of the asset,
   c. the minimum and typical batch sizes processed by the asset, and
   d. the changeover time required changing between products.
3. Draw, at the bottom of a landscape page of paper, these stages and their associated data boxes.
4. Identify the typical customer orders, schedules and delivery patterns in the top right-hand section of the map.
5. Identify the main supplier (key item of the bill of materials) and the typical amount of purchasing and delivery patterns in the top left.
6. In the middle at the top plot the order processing procedures and join the work schedule to the work station sequence of operations that have been written across the bottom of the chart.

Having written up the basic 'current state' performance map it is now possible to overlay problems and constraints of the 'as is' production system. You may find

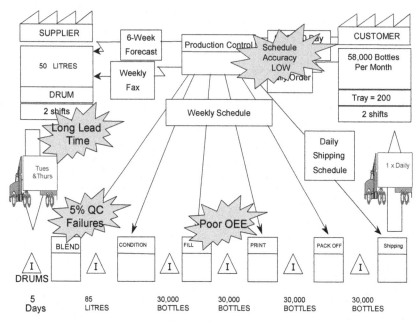

FIGURE 6.5 Problem map.

the use of a number of overhead projector acetates useful as a means of overlaying the problems onto the current state map.

An example of this is shown in Figure 6.5. It is not difficult to understand how the combination of poor OEE and quality problems will result in inconsistent delivery performance making it difficult to make accurate delivery promises to customers. These problems should be analysed, for every process stage and information channel, in terms of:

1. The process with the longest 'total cycle time' (the combination of the cycle time multiplied by the batch size and the time needed to changeover the asset) should be clearly identified. The process with the lowest average daily production output potential should also be identified. In most cases this will be the same process area but it can vary due to differences in shift patterns.

2. Problems affecting the quality of production at the stage in terms of physical losses.

3. Problems affecting the delivery of materials such as the availability of items, packaging or the lengths of the shifts in each area.

4. The cost per hour of operating the manufacturing stage showing the costs of using the work station to produce outputs.

5. Problems affecting the flexibility of the process stage to manufacture a variety of products including the identification of processes with long set-up times.

6. Now look for opportunities to improve the system through simplification, combination of activities that can feed each other directly and the elimination of activities that add no value (to the business or the customer) (Figure 6.5).

In parallel, the same analysis should be conducted for the top of the map and the information exchanges to identify what aspects of the order cycle can

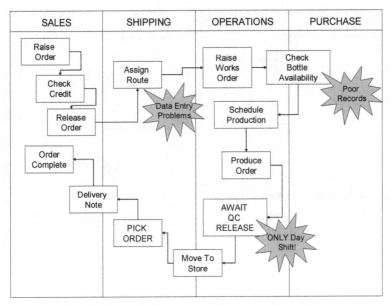

FIGURE 6.6 Administrative mapping using brown paper.

be eliminated or compressed to result in less lead time needed to get customer orders into the physical process of manufacturing. A recent study conducted by the Lean Enterprise Research Centre (Cardiff Business School) found this analysis to represent a significant source of loss and variation to automobile manufacturers (representing many weeks of activity but with just a day of time involved in the assembly of a vehicle (Holweg, 2001)). To conduct this analysis, it may be necessary to decompose the information flow to assess how orders are processed and through which departments. A brown paper map is a practical method for collating the outputs from such an exercise. This involves sketching out the information flows on a large sheet of brown paper and using this to record problems and wastes identified. In the example shown in Figure 6.6, the relationship between problems of data entry in shipping and the bottleneck of a single shift QC illustrates other contributors to the schedule adherence problems identified in the value stream map.

All too often, these hidden manufacturing processes represent procedures that have remained unchallenged even during the rise in popularity of quality circles. As such, many computerised planning systems contain obsolete data and processes. For instance, many manufacturing businesses update their computer planning systems weekly and as such have added a week to the lead times of all suppliers. On other occasions you may find that traditional safety stocks, entered when the system was being introduced, have rarely been questioned or that supplier lead times have changed but these have not been incorporated into the computer system data bank. As the Harvard Business Review article stated, managers must pin themselves to a customer order to understand what happens within their own system and the appalling number of hand-offs and delays experienced even before a production order is printed and given to the operations

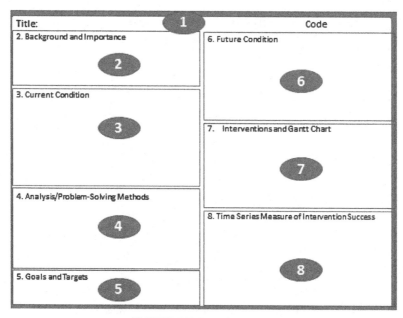

FIGURE 6.7 The A3 structure.

departments. You will be unpleasantly surprised to find out how long an order will wait before it is converted into an instruction to manufacture a product.

Armed with the ability to visualise the flow of value and draw a current state, the teams now have the ability to see where problems exist and how these problems should be reduced/eliminated to achieve the stabilisation of the production process. Value stream maps are also inextricably linked to the policy deployment process and the changes that will become management discussion and learning points.

Before we discuss more about stabilisation it is useful to introduce a method that helps to diagnose problems with production and incorporates the scientific approach of Deming's PDCA (Plan Do Check Act). It is called the A3 process. An example of an A3 process format is shown at Figure 6.7.

6.5 A3 LEARNING PROCESS

There is nothing special about the name A3 chart – it argued that it was simply the biggest piece of paper that (historically) could be faxed and read effectively by the fax receiver machine (a technology know pretty much replaced by email) – and that is why it became a useful standard document for communicating improvement programmes between factories at the time. An A3 is just a document that communicates the status of a project in a way that follows a systematic and scientific approach. The left-hand side of the A3 sized document contains all the problem-solving and root-cause analysis, and the right-hand side of the document contains all the solutions and implementation plan. And after an improvement is made the A3 document can be filed away to form an archive of improvements.

The A3 is the project charter of choice for many lean businesses because it represents the PDCA (Plan Do Check Act) chart on one piece of paper. An A3 document is written for every project – let us keep with the theme of 'Quick Changeovers' of equipment.

If we look at the left-hand side of the A3 chart (see Figure 6.7), it is possible to discern the following analytical steps:

1. The title of the improvement programme/project – the implementation of Quick Changeover with a certain set of bottleneck equipment.
2. Here the team lists the importance of the project and all the frustrations that require to be improved. In this case it would be excessive stocks of what is not selling and no stock of items that are selling. The frustrations could include that of the store team who constantly have to move materials, the finance department is frustrated because 'cash is sleeping' in the warehouse and not being turned into sales and such like.
3. At stage 3 of the chart, the team will often draw a value stream or process map of the current system with the addition of notes to show the issues that impact on the performance of the production and engineering changeover teams. These issues could include time losses through poor storage and retrieval of tooling, awaiting materials, constant adjustments of the equipment before the equipment produces outputs that meet specification and such like.
4. The mapping stage tends to involve a lot of perceptions and symptoms rather than root causes and tends to lack hard data so the fourth stage of the learning progress is to collect data on the size of the problem. This section makes use of the root-cause analysis tools of quality (Ishikawa, 1985) – so the question would be asked as to why does it take 4 h to change the equipment between batches? The team would identify the root-cause issues such as delays in finding the tooling, delays due to material supply to the equipment and such like. These types of loss would be counted and quantified (occurrences) and shown as a pareto chart.
5. In section five, the team determines the extent of improvement in the key measure of their success which we know is to reduce the average changeover time from 4 to 2 h and that the key measure is the number of minutes taken per changeover. The problem-solving part of the A3 is now complete.
6. The sixth box involves the team in redrawing – even cartooning – the future state and what the improved system will look like. In our example this could be (Figure 6.8). This shows a conceptual birds eye view of an ideal work area.
7. The next box contains the Gantt chart and implementation stages that have been designed by the team. As shown in Figure 6.9, this sets out the outline steps rather than the detailed task descriptions. The sequence of steps is unlikely to change which means that this chart can be used to assess the status and plan future steps in detail based on progress made.
8. The final box contains the key performance indicator (written down in box 5) which in our example is the average changeover time in minutes and this is equivalent to the bowling lane chart (typically as a time series chart) which has the same scale as the Gantt chart in box 7 – so you can see the

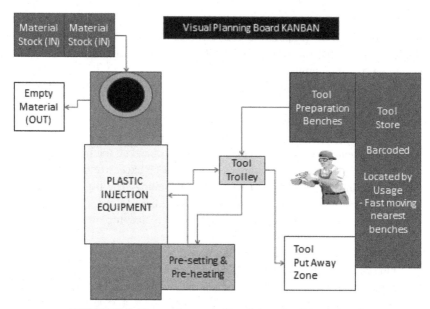

FIGURE 6.8 The new layout and flow (viewed as if from above).

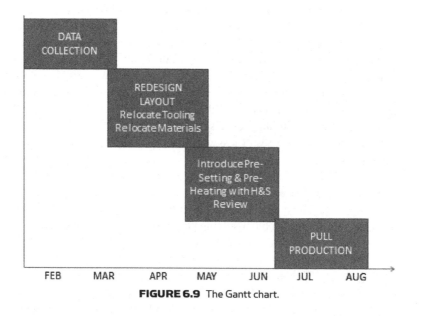

FIGURE 6.9 The Gantt chart.

improvement activity and its impact on the time it takes to changeover the equipment (Figure 6.10).

Once the A3 cycle is conducted and concluded then a final report is written which lists all the features of the change and lists any changes to the standard documents which may have happened during the change and the necessary approvals authorised (and documents updated to the new version so that the new system is compliant with its quality management system document reference).

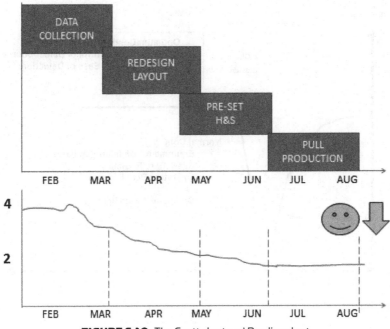

FIGURE 6.10 The Gantt chart and Bowling chart.

The A3 process therefore allows a full audit and a library of improvements to be created. It should also be noted that the value stream map can also be updated and even a future state of what ideal and optimised conditions of equipment performance look like – but more of that in the next chapter – this chapter is all about stabilisation.

Mapping the value stream and controlling limited improvement resources through the use of the A3 are important parts of the Lean TPM infrastructure and they are devices that confirm learning in a very powerful way. That journey begins with the restoration of the production system/equipment to a normal state (a major learning process in itself) because custom and practice and historical degradation of good practice are in sharp focus. Most employees will be thinking how the business ever let the production system and equipment decline to this state – hence the need to restore it.

6.6 THE ROUTE TO STABLE OPERATION AND ZERO BREAKDOWNS

The foundation of the route to stability is characterised by the TPM mantra of **'restore before improve'**. This principle should be applied to all 'best practice' operations, asset care, workplace organisation and administration activities as well as equipment. This the way to establish "Normal conditions" under which sudden sporadic failures no longer occur. As illustrated by Figure 6.11 progress towards this foundation step starts with getting equipment into a condition where it can do what is expected of it. To retain this condition requires

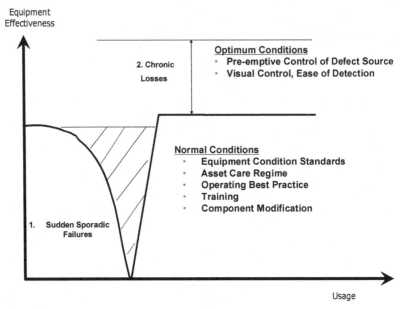

Equipment
Effectiveness

2. Chronic

Losses

Optimum Conditions
• **Pre-emptive Control of Defect Source**
• **Visual Control, Ease of Detection**

Normal Conditions
• **Equipment Condition Standards**
• **Asset Care Regime**
• **Operating Best Practice**
• **Training**
• **Component Modification**

1.   **Sudden Sporadic**
Failures

Usage

FIGURE 6.11 The route map to zero breakdowns and beyond. Addressing the causes of sudden sporadic failures, which results in chaotic conditions, is one of the first Lean TPM targets through learning how to set and maintain normal conditions.

compliance to asset care and correct operation best practice. Delivering these requires effective training and skill development. Only when these are in place should component modification be considered.

Managers need to start with the most critical equipment (as seen on the value stream map) and processes and must use the process of defining priorities to build ownership of all local managers and teams to the single change agenda. The change agenda has been developed by the policy deployment and catch-ball processes but it is still important to include production/maintenance and support groups in the discussion to encourage cross-functional rapport. The identification of the process bottleneck equipment is a key to making changes that improve performance (Hopp & Spearman, 2011).

It is important to use a simple scheme to assign critical priorities to equipment and processes such as:

• Category A = Critical all of the time/bottleneck;
• Category B = Critical some of the time;
• Category C = Normally not critical.

When doing this it is important to be sure to take into account the future product/business development plans for the next 12–24 months. This should present opportunities to integrate the Lean TPM process into the day job of all staff, and the highlighting of key equipment allows the team to see how important (commercially) the asset is in paying the bills and satisfying customers. Teams should also review existing improvement/capital plans and reassign investment priorities/resources. The plans and investments need to be separated out into improvement priorities that are high/low benefit and high/low cost. Priority should be given to low-cost/high-benefit activities. In many cases, the insights gained from low-cost improvements remove or radically change the scope of technical change required. For example, in a food processing company resources targeted at reducing labour intervention through high-cost automation of product assembly were diverted to introducing low-cost improvements in mixing processes.

Low-cost actions include minor refurbishment, the development of best practice standards and operator training. The outcome is more consistent product quality requiring less unplanned intervention during assembly. This released the level of labour targeted by the original project for almost zero capital investment. The benefit to the company and staff therefore comes from improved team working and increased confidence to address other hidden losses.

In short, the goal of this stage of the process is to take action to ensure that:

- equipment and processes are in a condition where they can do what is asked of them; and
- standards for best practice operation and asset care are in place.

These activities are usually sufficient to deliver 'best of the best' Overall Equipment Effectiveness (OEE) performance and typically this worth a 10–15% increase in productive capacity.

Some managers might question the value of such a simplistic approach and suggest the need to collect data to justify the effort. Indeed many six sigma quality improvement experts would like to collect a lot of data but this is not always necessary – such experts are much better employed to examine much more complicated issues. Experience shows that for the Lean TPM first step, when the actual costs are compared with the deliverables, the payback is typically measured in terms of months if not weeks. Furthermore, reducing breakdowns has an impact far beyond the immediate cost/benefit results. Put quite simply there is so much unnecessary waste and cost to remove from most businesses, it is best to avoid the risk that procrastination or 'paralysis by analysis' sets in and nothing happens to change and improve. The journey requires constant and small steps.

The true cost of breakdowns

Imagine yourself as a team leader on the shop floor when a breakdown occurs, how does it feel to have your plans for the day put on hold and then you have to reallocate resources? Such an event is never greeted warmly by your colleagues! And then the team has to spend the rest of the shift/week playing catch up. When such events are common, there is a 'repercussion ripple' throughout the value chain. This 'ripple' is the reason why spare capacity and buffers are designed in 'just in case'. Jobs are routinely brought forward to fill a gap in the schedule, materials are borrowed/stolen from other departments, equipment is cannibalised for spares, inventory and customer lead times are increased, overtime is called at short notice. The impact of these daily routine inconveniences is tremendous and working life can become a 'white knuckle ride' for managers and operations staff. It is a ride that soaks up precious management time, stops them from providing value (in terms of planning the change efforts) and loses production capacity that can never be recovered – and creates friction between staff as well.

TABLE 6.1 Firefighting and Time Management

| | | Urgent | |
		High	Low
Important	High	1. Firefighting	2. Proactive management
	Low	3. Poor prioritisation	4. Political

Table 6.1 sets out a matrix of tasks and those in box 1 are 'firefighting' tasks that must be acted upon immediately to restore a condition of normality. This normality is important because it provides a 'yardstick' for operations staff to detect changes and events that throw the production process into chaos and confusion. Think back again to the earlier discussion of 'demand amplification' – breakdowns and constant rescheduling aggravate this situation for your suppliers. When you constantly change orders this information is typically sent to suppliers, and guess what? They see your amplification and react to it. Such chaos is difficult to recover to normality and people become conditioned to thinking chaos is normal or worse stop complaining about it and just 'live with it'. This is a poor situation to face – living with it is no longer acceptable so concrete action to restore and then move to the future state is important.

If everyone in the factory were to 'blow a whistle' each time things went wrong, then ear defenders would be a compulsory safety wear at every factory – the trick is to make this whistle blowing the prompt for employees to engage proper and long-term countermeasures (and lower the noise!).

The root cause of management firefighting is, sadly, ignoring important tasks such as training and asset care – that is until they can no longer be ignored (they become urgent). A significant difference between world class and average companies is the amount of time that is spent in 'box 1' for traditional businesses whereas very little is expended here by the world class. For some individuals this is their preferred way of working, a macho environment, and one they have grown accustomed to – indeed some even enjoy it as it provides an adrenaline boost for them and has been the foundation of their career (the factory 'super hero'). These activities are nonvalue added, they are wastes, and they add precious little to the credibility of the company as a 'world class' manufacturer. They are incorrect behaviours and totally unsuited to the modern empowered workplace and they do not fool customers over the medium term because firefighting inevitably has an impact on customer service levels (in terms of QDC performance of the firm).

Constant interruptions to material flow require a series of improvements – there is no single solution or magic cure. Instead the level of firefighting dictates the quantity level of simple solutions needed to recover stability. In this respect it is worth considering: What would it be worth to the organisations to achieve zero breakdowns?' and 'How would that feel?'. The impact of such a situation has an enormous positive impact upon profitability, business confidence to plan the future and the motivation of the workforce. So stability is not so boring after all and offers a better return on investment, simpler and stress-free jobs and an ability to make dependable promises to customers. These are just the basics of operating a business and it is ironic that many companies aspire to these levels of management control.

To understand why 'zero breakdowns' is a realistic and achievable goal it is important to understand a number of key concepts:

- Breakdown means the failure of a component and this is different from a stoppage as a result of a jam, trip or blockage;
- There are only two reasons for breakdowns (poor equipment condition or human error) both of these can be overcome if there is a will to do so through restoration and training initiatives;
- 'Zero breakdowns' does not mean zero maintenance, if the plant condition is sufficient to run without breakdowns for a planned production cycle, the zero breakdown level has been achieved;
- There are world class plans that run for 12 h or more with zero breakdowns and examples of businesses that have 'believed and achieved'.

Further, significant progress can be made towards 'Zero breakdowns' using low-cost or no cost refurbishment actions and much of the restoration/zero-loss improvement programme will involve labour hours and time rather than high capital budget expenditure. This time is available for managers who believe that the workforce are fixed costs and not variable costs (to be fired as soon as possible). The trick is to break the 'old mind-sets' which can only be achieved by providing people with the necessary skills and time to improve the operations. Over time, with management actions and support, the old mind-sets can be eroded to the point that culture has changed and the deafening noise of 'whistle blowing' will subside. This will be evidenced as the 'overhaul mentality', where parts are replaced just in case, is substituted for data and the working lives of assets extended beyond the point that they have depreciated to zero on the company accounts. Breakdowns are a significant contributory factor to 'box 1' activities – they cause confusion, panic and the adrenaline rush but need to be tamed by the two actions mentioned previously:

- Set and maintain adequate equipment/process condition standards;
- Train personnel/simplify routine tasks to reduce the risk of human error.

Both of these issues are 'box 2' tasks that require leadership in terms of standard setting and implementation by functional and cross-functional management. Establishing standards and making them a habit will not be easy as it breaks the customary practice of operations staff (creating pain). This process is eased significantly when the people who are expected to change have a role in the design of how the systems will be implemented and operated and how best to convince others to think and behave differently.

The gains to the organisation of eliminating these sources of production system 'noise' and chaos include improved consistency, quantity and quality of output, cross-functional teamwork involving learning about how best to self-manage and a reduction in safety risks.

Eliminating high levels of breakdown therefore creates an understanding of normality for both operators and maintainers, and it is this platform that increases the value of production exponentially. Such an approach means

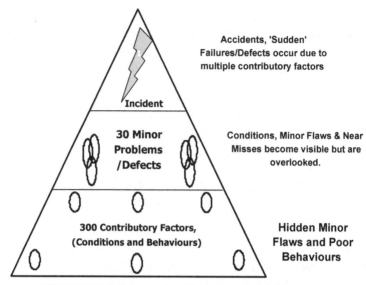

Accidents, 'Sudden' Failures/Defects occur due to multiple contributory factors

Incident

30 Minor Problems /Defects

Conditions, Minor Flaws & Near Misses become visible but are overlooked.

300 Contributory Factors, (Conditions and Behaviours)

Hidden Minor Flaws and Poor Behaviours

FIGURE 6.12 Patterns of accidents, breakdowns and defects.

establishing routine asset care and operations best practice 'as the way we do things around here' and enforcing these rules (taking disciplinary action where necessary for breaches). These routines, through rounds of team-based problem-solving, should be reduced to a minimal amount of time needed to conduct the tasks safely and efficiently (such as sequencing all inspections so they represent the least distance around a machine). Facilitators must challenge the teams to find innovative ways of shaving seconds from the process and finding new ways of recording that the tasks have been done (i.e. aircraft style check sheets).

As shown in Figure 6.12, where breakdowns occur frequently, this is a symptom of an operation which has many hidden minor problems and work-arounds. The fundamental root cause of breakdowns is poor or inconsistent management standards and priorities, typically the result of short-term cost down focus.

Weaning the organisation away from this outlook takes time but can be achieved by systematically applying the AIP (Asset Improvement Plan) template across the entire factory to set condition standards, formalise asset care and operating practices and refine those practices using focused improvement to systematically track down and eliminate the causes of breakdowns.

6.7 IMPROVING ASSET PERFORMANCE

The Asset Improvement Plan sets out a proven template of TPM tools to help management to kick the habit of firefighting and establish 'normal operating conditions' (see Figure 6.13 below). This incorporates a learning framework to provide the insight and underpinning knowledge to progress from problem-solving to problem prevention. It is also designed to be used in conjunction with lean tools to first stabilise and then optimise process flow potential.

Asset Improvement Plan

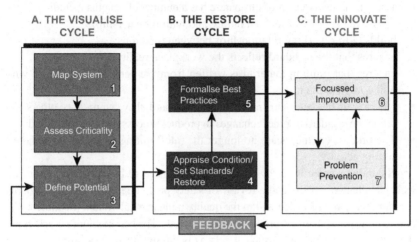

FIGURE 6.13 Asset Improvement Plan template.

The Asset Improvement Plan consists of three cycles.

The Visualise Cycle contains three steps to help the team to visualise the asset or system under review, how it should work and what its potential is.

The Restore Cycle guides the team in the assessment of asset condition standards and best practice for operating and maintaining the asset.

This includes the application of CANDO/5S techniques to improve workplace organisation, to remove the clutter to literally see the value adding process and the introduction of Single Point Lessons (SPL) to structure training and standardise best practices.

The Innovate Cycle these final two steps relate to the selection of appropriate focussed improvement and problem prevention tools including visual management to make tasks easy to do right, difficult to do wrong. Included in this cycle are tools to support innovation, transfer of lessons learnt and to lock in the gains.

The Asset Improvement Plan template provides cross-functional teams with a structured toolbox to stabilise and then optimise asset performance and deliver the
Lean TPM Zero ABCD goals:

- Accidents.
- Breakdowns.
- Contamination.
- Defects.

The route to delivering each of the ABCD 'zeroes' is raising work standards and making these a habit for staff. Only then can managers and team leaders be able to establish the correct behaviours. Where abnormalities are detected then these must be dealt with to maintain asset and material flow as well as basic staff discipline. The outputs from each project can then be translated into policy standards for transfer to other areas.

Ideally multidisciplinary shop-floor core teams will include technical specialists to ensure the quality of work conducted. This process will set the standards rather than process engineers on their own (the old way of imposing standards without team 'buy in'). Involvement of this nature has a number of benefits including:

- increasing shop floor understanding and with it ownership;
- builds closer working relationships between functions;
- ensures that the standard reflects the workplace reality;
- ensures that training material is written from the perspective of those who will complete the task;
- once a team has been shown how to set a standard, they can produce others and refine the standard to reflect changes in product/process specifications; and
- The process engineers are no longer the rate-limiting factor to the standard setting activity.

Naturally, standards set by shop-floor teams need to be subject to a suitable technical change process and a review of safety procedures before new standards are authorised and recorded within the quality management system (document reference numbers and review procedures established). The best medium for achieving the seamless introduction of such standards is through the use of an SPL (Single Point Lesson). An SPL is an A4 sheet of paper (with digital pictures showing good and bad practice) for operators to use like a motorcycle 'Haynes manual' (a user friendly publication for maintaining a specific motorcycle). An SPL therefore controls a process or task and is a reference document that is invaluable as a source of knowledge and standardisation which 'world class' businesses operate extensively.

6.8 LEADING THE IMPLEMENTATION OF STANDARDS

The outputs from the Asset Improvement Plan need to be captured at an asset level and used to support the development of enhanced ways of working.

These outputs, shown in Figures 6.14 provide a PFEA (Plan for Every Asset) and should be routinely updated as the asset performance progresses from stabilise to optimise.

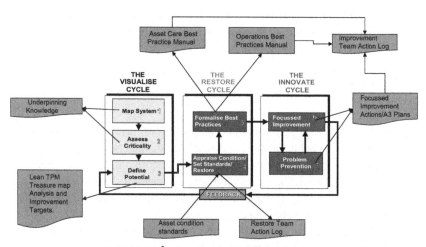

FIGURE 6.14 Asset Improvement Plan outputs.

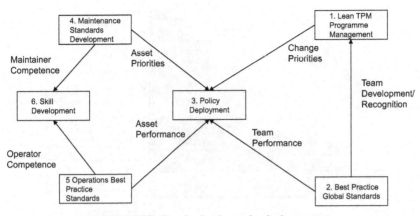

FIGURE 6.15 Standardisation underpinning systems.

Traditionally, in hierarchical organisations, shop-floor workers have been conditioned to receiving change rather than taking ownership for improvements. Supervisors have been conditioned to become co-ordinators of the by shift/daily 'white knuckle' ride of unreliability. Supervisors also used to make sure that workers are occupied, materials are available for the next run, arranging overtime and keeping the mass production system churning and people busy (even if they had no actual customer orders to complete! There was always stock to build!). Improvement in this environment is driven by initiatives and an elite of staff often involving 10% or less of the workforce – you would recognise these individuals because they are involved in all projects. Developing standard practices to support the development of production and maintenance capabilities is at the heart of the Lean TPM stabilisation process. As set out in Figure 6.15, the Lean TPM master plan drives the policy deployment process to challenge this traditional supervisory role. Changing it to a people development role in a very practical way. This is a powerful lever to engage all workers with proactive improvement and establish underpinning systems to support standardisation of working methods.

To guide the progress of both supervisors or more correctly first-Line managers and their teams, Lean TPM makes use of team development benchmarks (Figure 6.16). These benchmarks are progressive, when a team has achieved level 1A they are coached to deliver Level 1B. These benchmarks also provide the opportunity to reinforce behaviours and engagement by providing recognition at each level. Figure 6.17 illustrates how the Asset Improvement Plan template, implementation process and team development benchmarks combine to provide a transformation framework to raise equipment performance and provide a route map to high performance team working.

Incorporated into the Lean TPM team development benchmarks is one of the most powerful behavioural change tools known as CANDO, 5S[1] or simply

---

[1] It is called 5S after the five Japanese words used to describe each step of the workplace organisation process. The letters 'CAN DO' spell out the first letters of English terms for the same process. (Cleaning, Arrangement, Neatness, Discipline Order). Although these are not faithful translations of the Japanese words they are a useful acronym of the process steps.

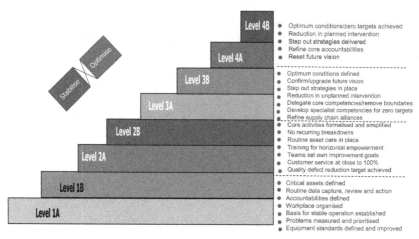

FIGURE 6.16 Lean TPM team performance benchmarks.

workplace organisation. Although often seen as merely a form of housekeeping, CANDO provides a simple route map to help teams to take ownership of their workplace and learn how to establish common work standards (see Table 6.2). In the table, the corresponding 5S term is included for those interested in understanding the origins of this thinking.

The CANDO steps can be applied to all manner of workplaces. It can even be used to set rules and routines for keeping your computer workspace (hard disk) clear of unnecessary items. Working with teams to introduce CANDO re-enforces a number of positive behaviours:

- How to establish rules/standards which the team will abide by.
- Empowering the team to take control of their environment.
- Setting standards which make it easier to produce good work.

On study tours to exemplar sites, the most frequently asked question is what you would do differently if you could start again? The most frequent answer to that question is that they would spend more time on CANDO/5S workplace organisation activities. The CANDO process is therefore an important stage and the state of the factory workplace discipline is a good indicator (and measure) of factory morale. When morale is low there is stress to maintain discipline and when high the factory should sparkle as a showroom to customers who may visit. Keeping the factory in a 'battle ready' condition should also be subject to rounds of problem-solving, simplification and reducing the time needed for such cleaning activity. For many businesses, especially those without a formal process of problem-solving, the CANDO activity is a great way for employees to learn about problem-solving and innovation before being taught about it and applied to machinery and processes. In effect, the CANDO activity uses a latent ability to solve problems concerning an issue that is close to the hearts of all employees. It should be noted that the close relationship between CANDO activity and workplace safety makes it a change initiative where early trade union involvement (in management and auditing of performance) is desirable. Safety is foremost in trade union thinking and developing the change

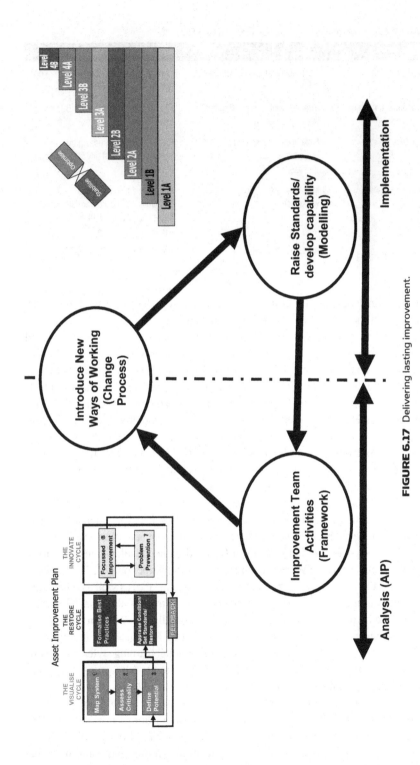

FIGURE 6.17 Delivering lasting improvement.

TABLE 6.2 Workplace Organisation (5S and CANDO)

Step	Goal	Improvement Theme
Cleaning (Seiri)	Get rid of what you do not need	Nothing in the workplace which we do not use every month
Arrangement (Seiton)	Arrange what is left so that it is where you need it	Find everything you need in 30 s or less
Neatness (Seisi)	Make it easy to keep the workplace clean.	Fix the routine for keeping the workplace neat and tidy with minimum effort
Discipline (Seiketsu)	Maintaining a spotless workplace	Understand how to reduce contamination/nonconformance and make it easy to do right (keep to the standard)
Order (Shitsuke)	Apply visual controls	Make it difficult to do wrong

mandate to include a role for the trade union is well worth having as it reinforces the legitimacy of the programme and assists with its policing.

To create a robust system of stable production operations, factory teams need to ensure there is effective training of staff to 'do the job right' – be it CANDO or any other stage of Lean TPM. Staff must also have the skills to 'do it better' as well as having daily routines that allow skills to be practised and daily controls to detect abnormality quickly (plus respond to it). The most effective design is one that is layered. A layered approach is a system whereby daily simple controls are layered with a managerial review process as well to checking activities have been done effectively. The routinisation of daily management is critical as we have seen – it is this set of planned routines that then allow for greater time for project improvement and policy deployment work.

Training Within Industry (TWI) is a systematic approach to training staff and creating the conditions whereby learning and practice result in standardised and effective routines to perform operations and management activities. The TWI process, which originated during World War II, was designed to allow conscripts to be trained effectively. The TWI approach includes Job Instruction (JI), Job Methods (JM), Job Relations (JR) and Programme Development (PD) as well as other additional programmes of training and evaluation. JI is a structured approach to teaching staff the correct method and observing their compliance, JM is an approach to understanding efficiency and making improvements, JR is a means of treating each learner as an individual and appealing to their individual learning styles and finally, PD is the means of spreading improvement methods to the full factory-wide level. For a full review of the processes see Graupp and Wrona (2010).

A lack of training (and regular review of competence) is one of the major reasons for failure in the factory and a barrier to stabilisation of the system. This chapter concerns the first two milestones of the Lean TPM master plan involving the focus of all employees on 'taming' technology, improving internal value streams and establishing effective cross-functional teamwork. The achievement of this foundation of robust performance will, as explored in the next chapter, allow the process of value stream optimisation and innovation to be exploited properly and with commercial effect in terms of generating a means of market advantage (Slack, 1991; Hopp & Spearman, 2011; Schmenner, 2012).

6.9 ESTABLISHING OPERATOR ASSET CARE

The table below (Table 6.3) sets out how the CANDO steps become the foundation for implementing first four steps of autonomous maintenance[2]. Although this concept is commonly shown with seven steps, the first four support the goal of zero breakdowns. The remaining three autonomous maintenance steps supports the achievement of optimum conditions and are dealt with in Chapter 7.

Achieving and maintaining 'zero breakdowns' is possible but not easy, it requires constant vigilance. With business routinely introducing new products and equipment, requiring greater flexibility and higher levels of it is easy for standards to slip. Like the golf pro who practises their swing daily, the maintenance of normal conditions takes discipline and a true commitment to high standards. CANDO provides the foundation for that discipline and commitment. What is more, if standards start to slip, you can see it. If visible standards start to slip, imagine what is happening to the discipline for processes that are less easy to observe.

Achieving normal conditions provides a predictable pattern of output. Although zero breakdowns is a 2–3 year goal, the benefits can be felt at bottleneck processes in less than 6 months. What you get is increased capacity and

TABLE 6.3 CANDO for Equipment and Processes

Step	Activity	Improvement Theme
Cleaning	Clean the equipment and restore functionality	Cleaning is inspection. To see what is happening
Arrange routine tasks	Formalise procedures and countermeasures to common problems	Detect problems and understand the equipment principles and improvement process
Neatness	Set cleaning and lubrication standards	Fix the routine for keeping the equipment neat and tidy and in good working order with minimum effort
Discipline	General inspection	Understand how to reduce contamination/nonconformance and make it easy to do right (keep to the standard)
Order	Apply visual controls	Make it difficult to do wrong

---

[2] The Autonomous maintenance is one of the core TPM principles and was developed by JIPM.

quality consistency. If the additional output can be sold, unit costs will improve. If the additional capacity cannot be sold, on its own, the benefits are significantly reduced. How should such potential be harnessed so that it is translated into value that the customer is prepared to pay for?

6.10 THE PROCESS OF STABILISATION: THE FREE-FLOWING MATERIALS MAP

The goal of stabilisation is to ensure work flows, just like a liquid through a pipeline, across the internal value stream and then through supplier companies in order to service customers. To understand the 'flow performance' of the current state production system, apply the OEE measure to each process stage and record the result on the map itself. The OEE calculations at the bottleneck operation dictates effectiveness of the entire value stream. It is this door-to-door OEE which will set the focus for improvement activities. It should again be noted that to focus all activities upon the improvement of this OEE measure has commercial benefits but to improve this figure may necessitate improvements at other points in the value stream. The logic is quite plain to see:

1. The bottleneck operation determines the output performance (what the customer sees in terms of the sequence and availability of outputs).
2. It is necessary to ensure that all work stations after the bottleneck do not generate quality losses (as these are very expensive and have used bottleneck capacity which is the most limited resource).
3. The sequence of materials getting to the bottleneck is important because this determines the availability of work at the bottleneck. Thus if the bottleneck output sequence determines the time at which goods are finished it is important that pre-bottleneck process has good quality and internal due-date performance and manufacture in a batch size that closely matches that of the bottleneck.

If a 'total production system' approach is taken to manufacturing and logistics improvement then all processes should be targeted with improvements – if only to ensure that they do not become a bottleneck. For example, an investment decision to buy a second asset in the bottleneck area may push the new bottleneck process to another part of the factory at which it will be difficult to improve unless work has already begun to achieve a zero-loss status at all nonbottlenecks. Think about it – it makes commercial sense to improve performance everywhere as any suboptimal performance does in fact offer no true value to customers but does add costs. It is important to remember though that the bottleneck process does have a major impact upon the commercial performance of the firm and also the flow of materials within it. As such, bottleneck management will dominate the design of the production system during the debate on how to get production flowing.

The Lean TPM treasure map with its 18 losses as set out in Chapter 4 will provide an insight into the wide variety of wastes that exist within the entire value chain. It is also worth calculating a few additional measures of performance. These include:

1. Multiplying together the average quality and delivery performance of the firm (both expressed as percentages) to its customers to the day stated by the customer. If quality performance was 98% and delivery on time was 70% then

this would yield 98% times 70% or a performance figure of 68.6%. Now the average number of finished goods held (in hours) should be calculated.

2. The internal OEE of the production areas.
3. The supplier performance should be measured in the same manner as the firm's performance to the customer and in the same time demands (i.e. to the day of requirement). Here performance is again determined as the average quality performance multiplied by the average delivery performance (both expressed as percentages). Thus if quality performance is 92% with 60% on-time performance then supplier performance to yield is 55.2%.

This form of diagnostic analysis really cements an understanding of a production system which each manager works within, makes decisions but often suboptimising the production system by optimising each business function and forgetting about the value stream. It is the stability and improvement of the value chain that provides the 'customer qualification' outputs needed by customers. It is also possible that this form of analysis yields the conclusion that free flow of materials in the current state is not possible and that the production system is underbuffered. The latter is usually the result of many round of pure internal cost cutting rather than system design and management using the key processes of QDC performance. Now it would be naïve of us to argue or propose that the production system can simply be changed in a short time frame and without problem. This would not be a good reflection of what the management of the factory has been doing until this point of attempting to optimise the value chain, would it? So practically there may be some solutions that still need to be found. In this case it is important to still continue with the improvement planning but revise certain aspects of the production system design.

To assure a good level of flow, and where it is physically possible, any production stage that is followed by another with a shorter overall cycle time should be combined. Similarly where flow is not possible, due to a number of reasons that will have surfaced during the mapping of the value stream, is to determine the points in the production system at which an amount of controlled inventory is needed to buffer the system. These buffers serve to protect the production system from disruption, the impact of high variety, machinery that constitutes a system bottleneck or other constraint. The buffers (investments of company cash in terms of stocks) are used to allow work to be pulled by demand and for manufacturing businesses to collapse their lead times to customers. In reality these buffers are quite financially cheap compared to the performance of the overall value stream and have the impact of slowing stock turns from an accounting point of view but commercially strengthen and support flow performance throughout the value stream. In short, costs will not be minimised but the quality and delivery performance of the production system will benefit. If introduced the buffers allow the flow of products (as measured in terms of OEE performance) to be stabilised, and this approach allows the factory management to revisit the need for buffers at a later stage. If you prefer, this approach is similar to using safety stocks to cover against problems whilst they are sorted out and then, at a later stage the safety stocks are removed without risking the performance of the system.

At this point the real improvements to the production system can be understood and the process for optimisation can take shape. These debates will tend to include the feasibility (given the amount of stock keeping units of the company) of using:

1. A total push system whereby orders are loaded and progressed through the manufacturing facility via the use of a schedule.
2. A total pull system involving the use of deliberate buffers of standard product to allow the finished goods of the firm to be lowered until a set amount is reached and internal re-orders and replenishments of the 'like' product are made. This fills back up the finished goods available for sale. As such, finished goods pull like products from the finishing operations and so on to the initial processes of the production system.
3. A hybrid approach is used where pull is introduced for production centres of families with low product variety or buffer spaces are used and withdrawals trigger the release of the next job in the manufacturing schedule. The latter is often associated with the management of the bottleneck operation and the subservience of all other manufacturing process stages to the schedule of this process. It should be noted that the bottleneck determines the overall capacity of the production system. The hybrid system therefore seeks to protect the operation with a buffer of either standard or scheduled products. An hour lost here can never be recovered and is lost to the whole production system whereas an hour lost at a nonbottleneck can be recovered as the nonbottleneck has the ability to produce more products quicker than the bottleneck.

Finally, once the basic production system has been created and stability restored within the factory, improvement activities are needed to maintain high performance as well as ever increasing the capability of the asset to result in flow without buffers or waste. The latter is reached when the bottleneck OEE has risen and stabilised at a consistently high level of performance, and the target for improvement is the 'door-to-door' OEE measure.

These analyses will, just as during the 'VOC' debate, unearth a whole host of issues that impact upon flow of materials to meet customer needs. These debates aimed at reconciling the design of the production system and the target performances needed to 'qualify' and 'win' orders over the next 3–5 years will include some suggestions for improvements. Starting with the internal value stream (within your factory) and the current state map produced in the previous section, it is possible to identify and cost the major projects that will release improvements in efficiency to increase the effectiveness of the firm and its ability to meet customer qualification criteria. The best way to achieve this form of systematic change and improvement is to combine an approach to cost deployment with an X-chart approach to justifying the expenditure of investments on production system improvements. These two powerful tools, used to quantify and focus the improvement effort, are invaluable when it is necessary to justify the use of limited business funds and to allow all managers to 'buy into' when the money is necessary. Such an approach is not typical of existing Western manufacturers who tend not to think about production as a system and often attempt to improve little bits of the production system that have absolutely no impact on the flow of materials through the bottleneck process.

6.11 LOCKING IN THE RECIPE FOR LOW-INVENTORY, HIGH-FLOW OPERATION DELIVERING ZERO BREAKDOWNS AND SELF-MANAGED TEAMWORK

Master plan milestone 1: all involved

So far in this chapter we have explored separately the bottom-up tasks of establishing the recipe for zero breakdowns and mapping/stabilising the value stream. In reality both activities are integrated. The value stream mapping activity will highlight critical and bottleneck processes. The improvement in reliability will enable the removal of buffers and revision of planning standards.

The policy deployment process combined with the roll out cascade introduced in Chapter 5 defines the programme and process to deliver the gains from this first milestone. Despite the need to deliver benefits, the exit criteria for the first master plan milestone are:

- All personnel are actively involved in the improvement programme and are working in improvement teams capable of achieving level 1A Team Review and Coaching (TRAC) benchmarks.
- First-line management ownership is beyond question.
- The gaps to deliver stable operation have been identified and a clear sequence of actions under way to deliver that.

Progress towards milestone 1 exit criteria is measured by combining the evidence-based TRAC scores at a departmental and site-wide basis. The trend of results is used as an input to the policy deployment process to confirming the glide path to achievement of master plan milestone 1 (all involved). This is also used to mobilise activities to progress onto milestone 2: refine best practice.

Master plan milestone 2: refine best practice

The title of this section sets out the goal for the second master plan milestone. The same techniques are applied as in the first milestone but as confidence grows and capabilities improve. As routines are formalised and simplified so that they can be delegated to release management and specialist time to lock in the gains and redirect their attentions towards higher value adding activities.

For this reason milestone 2 is characterised by the transfer of roles and upskilling of the front line team (see Table 6.4).

This part of the master plan begins the process of transferring routine activities to the operations team. This is a process which continues through the remaining master plan milestones as processes are optimised and the need for unplanned intervention to assure quality is reduced. The TRAC development process is also a high-performing team development route map which means that ultimately all routine activities can be carried out by the operations team. This means that they also become accountable for delivering the 'zero breakdowns' goal because it is within their capability to deliver it. As the capability to progress past each step is demonstrated, the TRAC process also provides a basis for team-based recognition to recognise and reinforce proactive

TABLE 6.4 Horizontal Empowerment

Current Role	Future Role
Operator	Technician
Maintainer	Engineer
Supervisor	First-line manager
Manager	Entrepreneur/Listener/Coach

behaviours. Despite this emphasis on the operations team, the progress will only be achieved if:

• the management provide the structure, systems and organisation;
• team leaders provide discipline, consistency and manage results not tasks (feed forward);
• teams have the capability, openness and goal clarity.

To achieve this level of management competence also requires the systematic development of those in specialist and management roles. This generally involves the development of skills to look beyond short-term planning horizons to be able to anticipate issues, prevent problems and manage the future. A key organisational capability to develop during this milestone is the ability to deliver projects so that they achieve flawless operation from day 1. To achieve this, organisations need to be able to enhance the effectiveness of their cross-functional middle and senior management level teams. This early management goal and the route to achieving it are explored in the next chapter on optimisation.

Exit criteria for this milestone 2 refine best practice are:

• Use of feed-forward mechanisms rather than feedback.
• No recurring problems.
• Stable delivery lead times.
• Routine tasks are carried out by self-managed teams/cells.

6.12 CHAPTER SUMMARY

Process stabilisation is the first significant milestone on the route to Value Stream Perfection (Womack & Jones, 1996). Achieving a 'restored normality' clears the technology landscape of the debris of poorly designed systems and working practices that lead to inevitable firefighting and customer service failure which act to ransom customer service. These first steps are as much about dealing with gaps in management processes and priorities as it is about innovation and smart technology. It is about leadership in terms of consistent direction/priorities, raising standards and releasing potential, and it is about formally writing down processes and procedures so that everyone can understand the system. In truth, most manufacturing systems are the amalgam of lots of peoples activities and specialist knowledge but it is only when you chart what happens on a day-to-day basis that you understand the rate (or lack) of flow. Putting all these issues on a single map helps highlight the issues – much of which is 'low-hanging fruit' and needs only a time investment. Just imagine what an organisation can achieve when all this firefighting is sorted and the exception rather than the 'norm'. For companies where every day is a 'white knuckle ride' this may seem a far fetched and unachievable but in 'world class' companies. It should be noted that this is not the ultimate goal but just this is only the entry ticket. The use of a current state and future state map is priceless in this process of visioning the future factory and showing employees the damage caused by processes that are out of control. Further still, it will awaken a commercial thinking that traditionally has not been part of workforce involvement. Imagine the horror as a machine operator, who experiences hassle on a daily basis to maintain performance from the machine, discovering that the company has 3 months of finished goods stock of a major item. This awakening is ideal and inevitably creates a sense amongst employees that *'if this was my business then I would run it differently'*.

When you have achieved this level of 'restored' understanding then the battle for 'world class' performance is underway. The hearts and minds of staff are now aligned in an effective manner, and value stream maps show how supply and logistics systems fit in to enhance cash flow and customer value. This scenario is far removed from the traditional factory systems where an operator arrives at work for another mundane session of 8 h moving bits without thought. Now you face a workforce who will question and innovate. Most managers will find this, at first a bit daunting, even threatening to their management practices but this will 'wear off' as this intent is constructive critical rather than personal. After all managers work within the system and try to do the best they can just like every other employee. The future state map should therefore be designed to allow managers to manage the big business issues and work with other managers whilst the shop-floor teams take charge of what they are best at – producing and improving the factory's material flow systems.

As we have seen, the first stages of Lean TPM clear up the noises and errors that interrupt product flow (and disruption to the process of banking the cash from sales).

This part of the programme concentrated on the first two solutions from the TPM problem prevention hierarchy that was introduced in Chapter 2. This is shown as Table 6.5 for convenience. This establishes basic conditions and standardised operator and maintainer best practices. The process is enhanced by using TRAC to build cross-functional team capabilities that align with a business-led master plan. The outcome is vehicle to engage the workforce with the pursuit of flow, flexibility and focussed improvement as a stepwise route map to industry leading performance.

The Lean TPM process of stabilisation dampens the impact of reactive problems in the factory by using the collective capacity and innovation of all employees. The early stages clear away issues of missing standards, ineffective working practices and poor training to allow technical staff to engage properly with process optimisation. A stable production system has very low levels of interruption to material flow and it is at the end of this milestone the management of the factory can revisit finished goods safety stocks, lead times, service

TABLE 6.5 TPM Problem-Solving Hierarchy

	Solution	Format	Added Value
5	Apply major capex project	Improve customer value	Deliver step out products and services
4	Apply low-cost automation	Reduce nonvalue added activities	Improve flow and flexibility
3	Refine process control	Optimise process controls	Improve mean time between intervention and reduce defects
2	Improve a best practice	How to carry out an activity	Improve work methods and competencies
1	Define/refine a standard or policy guideline	When this happens, do that	Improve visibility of standard

level commitments and work-in-process to further reduce the barriers to releasing the full operational potential.

This foundation makes it possible to fully exploit the potential of new production systems, lean methods closer business integration. At this point, lean methods such as Takt time planning (using the average rate of customer demand – which is effectively the rate at which the customer comes and takes products from finished goods) to plan production. As such if the rate of consumption is one product every 3 min then that is the rate that all production processes must be able to making products. If the cycle time of each asset allows for this to be achieved then (and only with a reliable system) can this form of advanced production planning be engaged. Without a reliable system there is no ability to really plan at all and this renders a takt time approach impossible to achieve. It is at this level of stabilisation and control that prize-winning levels of performance are achieved and many world class performers have passed through this stage such as Volvo in Ghent.

Having gained a position of 'normality' by stabilising the production system the improvement effort will slow unless managers and facilitators take steps to embrace the next master plan milestone challenge. The challenge, during which the maintenance role switches from preventing downtime to preventing defects is important, is one where the success of the business is aligned with a growth strategy requiring operations to strive for the next level of competence. The new learning organisation, based on great technical skills, is one where the mean time between interventions is extended for both production and maintenance – there is less process interference because the production system is controlled properly. It is a business where operational capability is a fundamental feature of competitive advantage. Taking operations to the next level, to perfect the delivery of products/services, tame underpinning technologies and establish industry leading working practices is the challenge covered in the next chapter.

REFERENCES

Graupp, P., & Wrona, R. (2010). *Implementing TWI: Creating and managing a skills based culture*. New York: Productivity Press.

Hill, T. (1985). *Manufacturing strategy*. Basingstoke: MacMillan.

Holweg, M. (2001) PhD Thesis, Cardiff University.

Hopp, W., & Spearman, M. (2011). *Factory Physics* (3rd ed.). Waveland Press Inc.

Ishikawa, K. (1985). *What is Total Quality Control? The Japanese Way*. (D. J. Lu, Trans.). New Jersey: Prentice Hall.

Schmenner, R. (2012). *Swift even flow*. Cambridge: Cambridge University Press.

Slack, N. (1991). *Manufacturing advantage*. London: Mercury Press.

Womack, J., & Jones, D. (1996). *Lean thinking*. New York: Simon and Schushter.

Chapter Seven

Process Optimisation

7.1 INTRODUCTION TO THE CHALLENGE

In Chapter 6, we set out how the dual targets of 'zero breakdowns' and 'stable internal value streams' provide the entry ticket to world-class performance. At this point, although there is much to be congratulated, there is a risk of becoming the hare in the children's story of the race with the tortoise. The hare loses his lead due to complacency and arrogance. It got half way round the race but did not finish because it lost sight of its end-goal – we need to keep on our toes. It is important, therefore, to be confident we know how to apply policy deployment process, the development of 'future focused' value stream maps and the drive to optimise. Stabilising the production process is not enough to compete with industry leaders – after all it decayed last time to the point we had to restore it! It is a fragile state prone to failure as customer requirements evolve beyond the current performance of the business.

Organisations content with this stage typically look for short-term gains (labour savings) to improve current bottom-line performance. The problem is that this will inhibit their capacity to transform as customers' expectations evolve and markets shift. As can be seen in Figure 7.1, each of the three generic strategic intents have to meet significant commercial, operational and technical challenges to assure future success.

Each strategy requires resources to meet those challenges. Simplistic cost down tactics can remove that resource leaving the lean organisation actually in an anorexic condition. Worst still, the resulting environment is unlikely to retain the talent needed to meet the challenge of future markets. It should be noted that it is not possible to excel at all three strategic thrusts. The operational competencies for a high-volume, low-cost commodity business are very different from those for high flexibility, mass customisation service of those who compete on customer partnership. The technology and commercial marketing competences for each strategic approach are similarly diverse. Decide on the most powerful strategic intent (based on your goal of delivering customer value now and in the future) and build your organisation to excel at that.

Some managers may feel that potential future gains are not worth the effort but all systems degenerate when left to themselves (Lord Kelvin, second law of thermodynamics) – so we need to work on them. Without the drive for improvement, organisations become complacent. Small problems are left until they

Strategic Intent	Tactics	Commercial Focus	Operations Focus	Technical Focus
Commodity Product	'Best value for money', low costs and scale	Multiple customers at 'internet' catalogue level	Minimise overheads, 'good enough' value at low price	Optimise operations, comprehensive range
Product/Brand Leadership	'Best product', innovation and speed	Brand Marketing, Promote end Customer Pull	Robust supply chain, excellent quality control	Optimise Product development cycle
Customer Partnership	'Best customer solution'	Understand customer value to define step out features	Operational Flexibility, able to bespoke product/ service features	Technical competence, to influence customer design brief

Choose where you need to excel to thrive in your chosen market

FIGURE 7.1 Generic strategic intent.

grow into large issues. Without the resources to deal with them, standards slip progressively resulting in death by a thousand small cuts. The predictable outcome is a downward cycle towards reactive management and we all know how painful that model of management feels.

Alternatively, by taking continuous improvement to the next level, the discipline of daily management continues to remove layers of waste which not only reduces costs but makes new opportunities visible to

- increase capacity at relatively low cost;
- improve quality consistency/reduce variability and the risk of human error;
- get closer to the customer and
- raise skills and with it the motivation of the workforce.

Avoiding the management decay curve and fatigue is important. Despite the significant challenges of the Optimisation milestones, they offer the best long-term gains – especially for managers. The achievement of these milestones also puts the business into a new league and one where the achievement of world recognised awards (such as the Shingo Prize and others) are achievable if the business seeks to acknowledge its operational excellence and achievements to date.

The fundamental mechanism needed to sustain the improvement drive is a transformational switch from actions being driven by '**away** triggers' to actions driven by '**towards** triggers'.

- **Away** behavioural triggers generate actions in response to events such as breakdowns, lost time, customer service failures or quality rejects. Although these are useful triggers to drive progress towards stabilisation, as that goal is approached, they lose their effectiveness.
- **Towards** behavioural triggers generate actions in response to the achievement of a future state. That is the delivery of a compelling business vision.

Working on the development of a compelling business vision certainly helps to reinvigorate progress towards the stabilised state and unite the teams in the

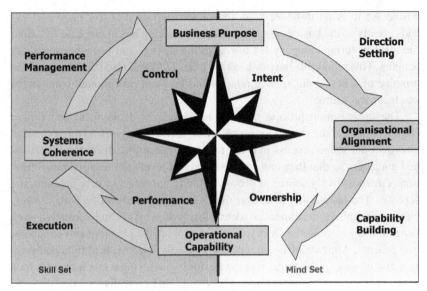

FIGURE 7.2 The strategic compass.

war on waste. The stabilisation state can still be achieved without a company-wide vision but, in contrast, the optimisation journey is doomed to failure unless it is driven by a strategic intent that has been translated into a clear, compelling vision and communicated across the organisation. The strategic compass (Figure 7.2) sets out the cause/effect linkages that translate strategic intent into strategic control. It also sets out the four skill sets which management need to master to apply this process.

Mastering Strategic Skill Sets.

1. Direction setting (pressure for change, innovation, goal alignment)
   a. Converting potential into business benefits and competitive capability through the future state through policy deployment.
2. Capability building (process improvement, value stream improvement, flexibility)
   a. Developing operational capabilities using lean (flow), process optimisation (TPM) to improve management processes (Lean TPM).
3. Execution (discipline, operations management)
   a. Engage all levels in improving operational execution so that it is disciplined and robust yet flexible to changes in demand and new product/service introduction.
4. Performance management (knowledge management)
   a. Learning from the past to improve future performance and deliver incremental profit improvement.

As shown in Figure 7.2, management processes 1 and 2 impact on mind-set and processes 3 and 4 on the development of skills. Through mastering these skills, managers in exemplar organisations are able to convert the gains delivered by front-line teams into bottom-line benefits by directing those efforts towards capabilities that add value. This might include possibilities for some businesses to bring in-house work that is subcontracted at this point and that adds greatly to the business and its future. Re-shoring has in one case of a floor tile manufacturing company led to staggering savings and further productivity benefits. This represents the pay back for Lean TPM and also the delivery of the promise of a new collaborative relationship between operational, commercial and technical teams.

The optimisation process drives a new cycle of deployment and further empowerment of the shop floor teams. More needs to be given to the teams that now own the stable process in order to release more time from the specialists and managers so that they can engage in greater levels of business transformation. The most likely source of growth is the introduction of new products and services. The fastest way to deliver that is through collaboration with strategic partners (suppliers, customers, academic institutions). This transformation work benefits from work in the Oobeya room to support the planning and collaboration process. After stabilisation, projects typically improve lead time compression and cross-organisational process flexibility involving a much broader look at the business and its supply chain. There will also be a realisation that the business must work only with companies in its supply chain that are committed to improvement and share a co-destiny in opening new markets and enhancing the product offering to customers. To reach full optimisation of the process, it will be necessary to engage the supply chain and spread best practice and that may imply teaching suppliers how to implement Lean TPM. Any aggregate improvement in supplier performance means that valuable space (currently racked out with stock) can be reduced and converted to a productive use for another product. During the optimisation stage, the processes of supply chain, time to market for new product designs and capital expenditure are critical and likely to be targeted for improvement.

Work on the new corporate vision and new rounds of policy deployment should have started during the stabilisation milestones but progress towards the optimisation milestones is dependent on mastery of the four management skills set out in the strategic compass above. This is the process which will absorb the experts whose time has been released from the day-to-day management task. They need to learn how to collaborate to lead the organisation into unchartered waters of step out products and services. Hence the reason to return to the Oobeya room with a new policy deployment challenge to solve! Milestone 3 of the master plan concerns mobilisation and engagement with the optimisation challenge. As mentioned earlier, this will depend on the organisation's ability to deliver projects flawlessly from day 1 (see early management (EM) below). Milestone 4 concerns the full realisation of the current optimisation challenge. It also includes the development of the next 2 milestones setting out the future 3–5 year vision to sustain progress past milestone 4. This is the mechanism by which exemplar organisations secure never-ending improvement.

This provides the route map to enhance the potential of competitive advantage for ever. This long-term perspective is essential if we expect suppliers to buy into the overall vision for growth. No-one ever said that creating and sustaining a lean business was easy and if it was then there would be no competitive advantage to be gained by doing so. It is no wonder then that only around 1% of companies achieve and sustain true 'world-class' levels of performance. These few have achieved the ability to set the trend of change in their chosen market and to do so in a manner that commands customer loyalty – and all this is based on enhancing value from the production, design, supply and all other elements of the business.

7.2 CHANGING MIND-SETS

Behind the difficulty of improving business performance for competitive advantage, through a manufacturing-led advantage, lie a number of major issues. Professor Yamashina, an internationally accepted and respected TPM guru, outlines these by proposing the following:

1. In the West, managers do not tend to train operational staff in the functionality or technical knowledge concerning correct operations of productive assets.
2. In this situation, mature continuous improvement activities have eliminated the low hanging fruit of poor materials supply, untidiness and lack of standardised work. Improvement teams may then identify problems/opportunities to improve assets but without basic engineering skills, they will not be able to resolve them.
3. The inability to solve engineering-related performance problems necessitates a greater integration with the engineering departments (training/support) which results in frustration for operational staff if these issues are left unresolved or poorly supported.

According to Yamashina, the result is a 'glass ceiling' for skills and improvement activities. Figure 7.3 illustrates how this glass ceiling prevents organisations from progressing past the stabilisation stage. This is characterised by organisations that take labour costs savings as they approach zero breakdowns rather than refocus the released management, engineering and specialist resources to raise their competitive capabilities. This is a serious issue and goes directly to the heart of developing a self-sustaining company-wide Lean TPM improvement programme. Application, application and application is the answer to sustaining change but what is important at this stage is defining the right questions to ask of the production process and ideal working conditions as much as asking questions about what the future employee should be capable of achieving.

There are a number of reasons why this 'glass ceiling' can be so difficult to breakthrough. Few of them are technical, most relate to human behaviour. The paragraphs below set out the key factors to be overcome.

Organisations which have become used to fire fighting, often relax their efforts as they approach zero breakdowns – they think the 'war on waste' has somehow finished. The feeling that the fire has been put out will lull the unwary into a very false sense of security. When management fire fighting is the norm, there is a certain satisfaction in getting through the day unscathed and even a dose of being an organisational hero who 'saves the day'. Such adrenaline and 'rush' is not conducive to making money and when you take the problem-driven adrenaline rush

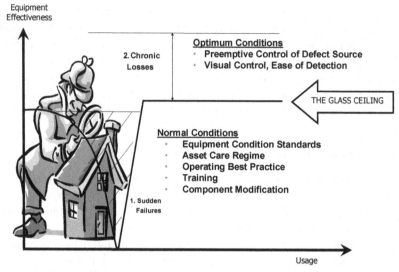

FIGURE 7.3 The route beyond zero breakdowns.

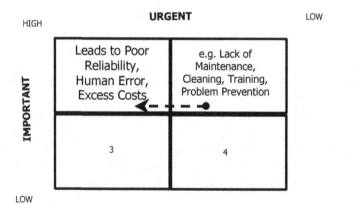

NOT ALL IMPORTANT TASKS ARE URGENT

BUT SOME WILL DRIVE BEHAVIOURS IF IGNORED

FIGURE 7.4 Behaviours reinforcing reactive management.

away then the working environment can seem very dull. It is usual for organisations which achieve stable operation to feel that order volumes have dropped when they are actually processing more business with less effort (one of the outcomes of the lean approach). To hold the gains, such businesses need to learn how to be driven by what is important rather than what is urgent and the fun that comes from being in control and having the time to think. The implications for the organisation can be far reaching. As shown in Figure 7.4, the application of maintenance is one of those important tasks which can become urgent if not addressed.

The application of maintenance can be likened to that of the gas supply in a hot air balloon. On the way up, the flame can be reduced and the balloon will keep rising. Once the balloon begins to fall, if the gas to the flame is not increased, it will reach a point after which the balloon will crash no matter how much gas is applied. This point of 'no return' is not obvious but it is none the less certain. Operating in the important/nonurgent (prevention) sector requires the pull of engagement with a compelling vision. If the organisation has not learnt how to achieve this, it will not progress past the glass ceiling.

Heightening the sensitivity of the front-line staff

Engagement can be quite a difficult skill for first-line managers to acquire after years of learning how to make do and mend. Another group whose role also changes at this time are maintainers (see Lean maintenance below). For the team leader, there are real benefits especially as 'time' becomes available to manage the medium-term business issues and engage in business development activities through cross-functional groups of other leaders. The challenge to optimise is a welcome one for the maintainer, if a little uncomfortable, in that it calls for the application of the 'technical skills' learnt at college. If this sounds exciting it is. If it sounds far-fetched, you should be aware that the two companies (mentioned above) had achieved this state and perform at extremely high levels of OEE (Overall Equipment Effectiveness) since 1997. Unfortunately, both are in Japan but could so easily have been located in the UK if businesses had engaged in Lean TPM sooner or some of the TPM pioneers in the UK had not given up or had their specialist facilitators poached by others.

Lean maintenance

Maintainers need to build on their knowledge of how to prevent failures with more about how the process works so that they are able to optimise the process. With less production system 'noise' that most precious resource is released to the maintainer – time and time to invest in improvements/projects. This evolves traditional maintenance into lean maintenance. That is, the traditional lean approach to maintenance is to classify it as a necessary but essentially nonvalue-added activity. The reality is that the act of restoring wear and tear secures future customer service. In addition, the experience of aircraft maintenance shows that effective maintenance minimises intervention. Within TPM the role of maintenance is to stabilise and extend component life. It could easily be argued that all of these activities add value. However, the bottleneck in most lean implementations is engineering resource. Also once zero breakdowns have been achieved and routine asset care is an integral part of operations best practice, the maintenance focus shifts from preventing breakdowns to preventing defects. This is where the real value-adding role of maintenance rests. It should also be noted that the limits to growth of whole economies is frequently attributed to the level of skills. Close to the top of that of missing skills list is usually a lack of engineering skills. The optimisation phase of Lean TPM releases the engineering capacity to add value. Under

	Maintenance Improvement Gains			Lean Manufacturing gains without Lean Maintenance	£k Total Lean Gains
	£k Improved asset care	£k Process optimisation	£k Impact of Lean Maintenance		
Plant 1	£108	£68	£176	£228	£404
Plant 2	£228	£157	£385	£456	£841
Plant 3	£313	£692	£1,005	£1,357	£2,362
Totals	£649	£917	£1,566	£2,041	£3,607
	41%	59%		57%	100%

1. Asset care/reliability improvement worth less than half of the Lean maintenance improvement gains

2. With Lean maintenance, Lean programmes delivered over a third more in bottom line gains.

FIGURE 7.5 Lean maintenance added value.

Lean Maintenance approach, the role of maintenance transforms from preventing downtime to preventing quality defects. Figure 7.5 illustrates the gains made by this change of emphasis. As can be seen, the benefit of the improved asset care to reduce downtime is significant but the gains from targeting the causes of quality defects added significantly to these gains. Further more together these gains had a significant impact on the overall benefits delivered from the Lean Manufacturing programme at each of the three sites. An important outcome of the Lean Maintenance focus on preventing defects is to extend time between intervention for both production and maintenance personnel. That provides a route map to higher levels of automation, capacity and productivity and flexibility.

To return to the earlier analogy, an organisation, which is skilled at repairing crashed hot air balloons, will tend to develop an organisation, systems, capability and culture which support this. To progress further, the organisation needs to develop new capabilities and a new management model. Milestones 1 and 2 provide the foundation to progress from fire fighting to a stimulus based on engagement with a compelling vision. In this chapter, we focus on the process of transformation from one successful business model to an as yet undefined future model.

The first leadership challenge is to change these and similar mind-sets. This is a collective challenge linked to the business case for raising standards, pressing further with the Lean TPM principle of customer-focus and engaging the passion to improve. Combining the winning business case with the Lean TPM programme incorporating the team review and coaching process provides the recipe for success by proactively creating the culture the business need, not leaving it to chance. Without the pressure of a winning business strategy, Lean TPM or any other improvement programme will not deliver lasting results.

The management challenge is to establish the hard-edged discipline to remove fixing and fiddling in favour of structured improvement based upon data and targeting. Truly 'world class' managers have this intolerance to a level where they

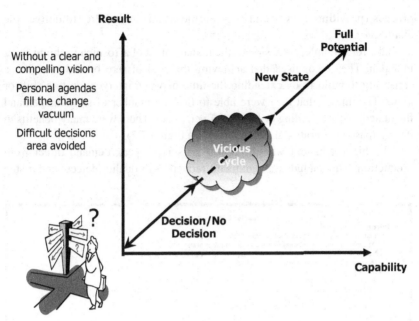

Result

Without a clear and
compelling vision

Personal agendas
fill the change
vacuum

Difficult decisions
area avoided

Full
Potential

New State

Vicious
Cycle

?

Decision/No
Decision

Capability

FIGURE 7.6 Breaking through the glass ceiling.

cannot hold back from exposing poor practices and standards which are not good enough – not for the traditional dispensation of blame but simply that these events are learning points for everyone. If your future state organisation has the belief that employee integration is a must, then you have no option but to subscribe to the 'no-blame' culture – it is the system that is wrong (and that is more often down to management!). Blame at this stage will be regarded by operations teams and others as a return to the 'old ways'. Further, these events and intolerance are not a sign of arrogance but simply the observations of 'trained eyes' that can see the value in removing waste in a process even though it is basically stable (issues that still need to be solved with permanent fixes will naturally be the focus of new A3 improvement efforts). Put simply the next two milestones shift the focus from problem solving to problem or opportunity finding in pursuit of customer/market leading capability.

What gets in the way?

In Chapter 2, we set out the route map followed by organisations that have successfully broken through this 'glass ceiling'. In each case, these organisations have been driven by the passion to develop and implement strategies which *require* higher levels of effectiveness and which have the attraction of delivering outstanding competitive advantage. This has been a key part of their breakthrough process. Not all of them made it at the first attempt. Figure 7.6 illustrates the dilemma of those initial failures.

These companies recognised that the passion for improvement feeds on challenge for factory teams and helps to accelerate the empowerment process (providing they are also given the means and ability to deliver the challenge).

Take for example, the vision of the management of a Toshiba Lighting Plant in Japan. They recognised that achieving the goal of zero breakdowns was a capability to build in by extending the time between intervention to a shift or more. This meant that they were able to flex the number of shifts as demand fluctuated, without adding or laying off personnel. There were many benefits to this increase in operational capabilities (see Figure 7.7).

The biggest benefit was the ability to service peak demand direct from production. This included a significant reduction in quality defects and a step

---

Toshiba Lighting Case Study

1.0 Prior to Optimisation
 It could take 3 years to get trained on some jobs
 Machine maintenance routinely required input from specialists
 Automated equipment took up too much manpower because of minor niggles
 There was a lot of waste due to set up and adjustment during change overs

2.0 Business Strategy
 By developing the ability to operate throughout the night shift without routine
 intervention, capacity could be flexed to meet seasonal demand without the need to take
 on additional (temporary labour)
 This reduced a major source of defects and associated costs
 It also broke the link between productivity and labour

2.1 Focussed Improvement Goals
 Eliminate waste
 Create an attractive plan full of vitality
 Develop capability to tame technology
 Zero breakdowns
 To achieve industry leading performance

3.0 Approach
 Technical Focus: Trapping defects at 19 auto inspection points
 Use this to gain insight into cause effect mechanisms and to eliminate the source of
 defects
 Safety focus: To abolish unsafe practices, to abolish unsafe areas, to improve discipline
 Environmental Focus : Reduce noice by 35%
 Working Methods: Refine and simplify working methods to reduce risk of human error,
 sources of defects and time to develop capabilities
 Skill development: Establish training categories (Core, Intermediate, Specialist) and
 support achievement of defined capability standards

4.0 Results

	Year 1	Year 4
Breakdowns per month	387	33
Productivity	100	247
OEE	71	88
Minor stops	100	11
Defect rate	1000	9
Customer claims per annum	89	8

4.1 Intangibles
 Sharing problems and successes
 Shop floor issues come first
 Equipment knowledge enhanced
 Workplace full of vitality (High Performance Teams)

4.2 Lessons for communication of design issues (EEM)

Source of Design Problem	Impediments to Communication
Operational details	Mismatch in education
Test/check lists	Depth and breadth
Fault diagnosis	Low designer motivation
Written details and drawings	User contempt
Ability to understand	Human inertia and procrastination
Capability to locate the fault	Presentation/form of documentation

5.0 Future goals
 Develop our own brains
 Implement improvements to achieve lower costs
 Enhance individual skills
 Improve profit and environmental benefits simultaneously

FIGURE 7.7 Toshiba lighting case study.

change in productivity because they no longer needed to hire and train temporary labour to man additional shifts. Again, the significance of this 'pursuit of perfection' can never be overstated. The result is a fitter and leaner business that is capable of opening new markets as much as it can now defend itself during 'price' competition (and without giving away valuable margin!).

A similar approach was used by a cement processing plant that had already made major improvements in zero breakdowns – the challenge was to double again the time between stoppages so that they could run with confidence through the night (and avoid having the burden of high overtime bills or the third shift for parts of a year). In addition, the completion of this challenge provided the 'business winning' ability to take customer orders up to end of the working day for next day delivery. This could hardly, given what we have said before, been seen as a boring challenge to keep people going – it was instead a significant addition to the competitive arsenal of the firm.

What distinguishes the managers in these companies from all others is their ability to think and act strategically and improve the **production system**, rather than being driven by short-term mechanistic cost reduction exercises and interfering in a piece meal manner. Such short-term thinking puts head-count reduction high on the list of cost-reduction targets which is illogical and counter to the beliefs of 'world-class' companies. World-class businesses, there will always be the need for people in transformation processes so they keep job descriptions 'loose' to allow redeployment of displaced workers within the factory processes and they rarely lay off because to do so can ransom the improvement process. The latter is an important issue, and if workers believe that their improvement ideas will harm the livelihood of co-workers then innovation will stop dead – 'no-one wants to improve a mate out of a job'.

Care and attention is therefore lavished upon the operator grade so that they can play an increasingly important role in the management, control and improvement of technology. Reductions in unplanned intervention means operators are not the 'output rate' limiting step and multiskills ensure people can move fluidly to where they are needed in the process to improve or stand-in for others. As such, to move to an optimal state of production performance, the operator is trained to take on new roles including

- routine production, maintenance, hygiene, environmental and quality assurance tasks;
- short-term planning and routine daily management administration;
- capture of lessons learnt and enhancement of best practice;
- setting and delivery of OEE improvement goals (within policy deployment);
- dealing with peers within customers and supplier organisations supporting
  - the introduction of new products/services and materials;
  - developing innovations to reduce total supply chain cost.

These are but a fraction of the opportunities for new role development available for businesses that have gained stability. Many choose to engage operators in additional qualifications and most businesses will seek out a greater relationship with a local technical college in order to devise 'company-specific' courses delivered to employees at the place of work. As the operator role matures, the role of the maintainer changes, away from fiddling and fixing breakdowns towards an engagement in optimising the process.

Springing the strategy trap

In summary, moving past the potential watershed of master plan milestone 2 depends on the development of a business vision and strategy which can inspire the whole business. One which can progressively replace the fire-fighting/**away** action triggers with ones which secure progress **towards** a collectively worthwhile goal. Many alternative futures could be available to those who achieve milestone 2 stable operations. Time taken to make a sound choice is time well spent.

Basing strategic thinking on past/current business models will reinforce the status quo. When the market shifts and the business model no longer delivers results, few clues will be offered about what to do next. The business graveyard is littered with examples of business models which once worked but then failed. Springing the strategy trap requires a different outlook. Figure 7.8 below sets out this importance of considering multiple futures to achieve a robust strategy. Past success is no guarantee of future success.

Examples of predictable external pressures include interest rate and currency movements which are quite predictable, what is not known is the timing of them. Research shows that 10–15% of customers will change supplier due to circumstances outside of their control and 45% of customers review their suppliers each year. It is predictable, therefore that some customers will be lost. Robust strategies are flexible to such changes because the lead team have

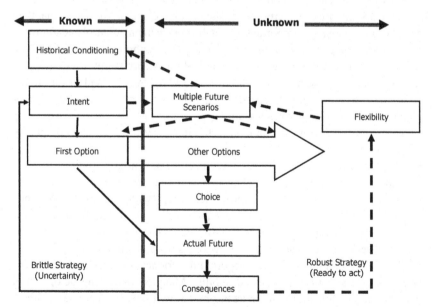

FIGURE 7.8 Springing the strategy trap.

considered their response to such events and can move quickly when they occur. This strategic assessment of business scenarios should not result in analysis paralysis but should raise awareness and interest in the suitability of available strategic options so that the organisation can keep these open as long as possible.

The importance of this deeper level of thinking ('the one business' approach) cannot be underestimated because without it, there remains the risk of a 'knee-jerk' response to a shifting market by cutting headcount (and slicing out employees with value-generating potential). As well as the loss of future opportunities, redundancy scenarios leave a lasting scar on those who work in the organisation and again reinforcing the view that management cannot be trusted or that they are incompetent and cannot be relied upon.

It is important to recognise that analysis will only take you so far. After all the options have been evaluated, the brutal truth is that when embarking on a new strategy, all of the details cannot be known. If this were not so, there would be no point in trying to compete.

Organisations will not break through the 'glass ceiling' without developing trust, mutual respect (that does not have to include liking management!) and a feeling that there is a co-destiny between the survival/growth of the firm and the individual employee. This is not to say that in certain circumstances headcount reduction is not appropriate but it is exercised fairly and undertaken to ensure the survival of the business or to compensate previous poor decisions have left the organisation vulnerable. Lean TPM is a tool to help management avoid such mistakes by changing mind-sets through a focus on business growth and engaging the workforce with meeting that challenge.

7.3 CHANGING SKILL SETS

The second two management processes within the strategic compass focus on the development of skills through practical application. A common barrier within traditionally structured hierarchical organisations is the way in which individual business functions operate in isolation. Often these functional silos will set their own strategies rather than being an integrated and aligned element of the business. Such artificial boundaries can generate internal conflict. Such poorly aligned businesses can be successful even though they retain an insular departmental focus (even keeping this view during milestones 1 and 2) but there comes a point when the lack of business process thinking and 'business level focus' creates problems with decision-making speed and the maintenance of these self-centred structures will limit the pace of change and improvement (Hammer, 1996). Reaching the point where these departments must be integrated is easy to detect, it is the point at which progress towards improvement goals stagnates and political infighting fills the vacuum. Here businesses will lose the good and mobile people unless a new focus is provided – a focus where customer value is maximised through collaborative internal projects to improve business performance.

A potential for change is therefore to build new mini-businesses within a business that is structured around a distinct value stream – value streams offer a focus and the Pareto rule of annual sales volumes adds a real customer to that mini business. Moving beyond the 'glass ceiling' depends on the evolution which supports such collaboration to secure horizontal empowerment. As mentioned previously in this book, this point occurs when routine processes are simplified and what was previously a specialist activity can become part of the organisations core competencies. For example, once equipment is in good condition and the activities needed to maintain this condition are standardised, Operators can carry out routine maintenance because they are closest to the process and with training can identify when intervention is needed. Eventually the operational team will develop the capability to deal with all routine activities areas such as planning, quality control and internal logistics. This is an important mechanism for releasing specialist capability to support delivery of the new strategy. The low level of interaction in hierarchical organisations inhibits this process. Figure 7.9 illustrates how manufacturing organisations can be defined by three broad functions.

Such optimisation can only be achieved where there is a deliberate policy to perpetually upskill workers (at all levels) and the approach of 'teaching a man to fish so that he can feed himself' is the approach that must be taken to skill development. This means that organisational structures need to be flexible enough to cope with blurring the edges of these overlapping functions to encourage higher levels of integration and support rapid change. However, this capability to involve indirect functions is not at the expense of removing these functions altogether. The release of routine activities to the operations team releases the skilled resource to reach new heights of business performance and the technical training that the TPM guru Professor Yamashina so cherishes.

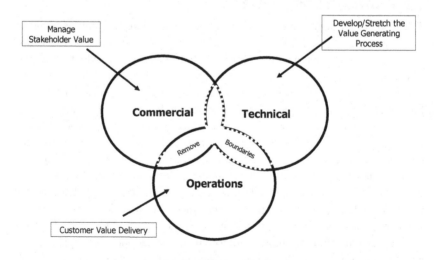

Representing the Sum Capability, Knowledge and Experience of Those Involved in Delivering Customer Value

FIGURE 7.9 Horizontal empowerment.

7.4 THE OPTIMISATION PROCESS

Milestone 3: build capability (extend process flow)

The aim of milestone 3 is to identify the recipe to release the full potential of the current operation and build the foundation to match and exceed future customer expectations.

This can be said to have been achieved when the organisation is fully engaged with the task of delivering customer leading capability, characterised by the following exit criteria:

- have identified critical optimisation targets (commercial, operations and technical) and making progress towards them;
- maintained zero breakdowns whilst transferring routine condition monitoring and servicing from maintenance to production personnel;
- established a clear product/service 'innovation stream' strategy in place capable of achieving customer leading performance;
- focus shifted from internal improvement to include external partners and
- technical focus shifts to external scanning for asset innovations rather than internal correction. Engineering staff to ensure that they are included in any future asset specifications at the capital expenditure and procurement stages. Feedback by improvement teams to this knowledge base is routine. This includes intelligence about improvements needed for the next generation of assets.

Top-down process

Given that the programme scope includes the leadership agenda, the generic management focus for approaching the next dimension of the Lean TPM 'perfection' Rubik cube is 'optimisation' to deliver multiple 'zero targets' (not just breakdowns but all forms of loss). Here, loss modelling and cost deployment tools are useful focussing activities. Loss modelling is used to identify the links between cost and hidden losses. It provides a top-level prioritisation process which is capable of considering complete supply chain optimisation. This is important as for example, activities to improve flexibility, shorten lead times and reduce WIP/finished goods inventory levels can result in increases in raw material stocks. Here suboptimisation of one part of the chain helps to deliver lower overall costs. Another example, the outsourcing of cleaning tasks may make sense in isolation but can increase maintenance costs and production downtime through the use of low-grade labour. Without the precursor of loss modelling, cost deployment can constrain supply chain optimisation.

As part of this review consider each of paths through the value generation hexagon map (Figure 4.4) taking into account current capabilities and future needs. Using the hidden loss treasure map (Chapter 4), assess the impact on business performance of the future state value stream map at 100% supply chain effectiveness. If this seems too fanciful, set what you believe to be a practical 3–5-year improvement goal (typically a doubling of current effectiveness levels). This assessment should include the benefits of increased sales volumes as well as reduced direct costs per sales unit.

Assess each area of the treasure map in terms of potential contribution using a simple scale 1 to scale 3 where 1 is no impact, 2 is some impact and 3 is significant impact. Use this assessment to apportion the estimate of overall benefit to the most significant areas of hidden loss. Allocate a champion to review each loss area in detail to identify current costs and potential benefits. Figure 7.10 shows a typical output from such an exercise.

In this way, the loss modelling helps to identify priorities for improvement and provides a key part of the cost deployment brief for an appropriate multi discipline team(s). It also provides the mechanism for setting goals and monitoring progress (typically quarterly) as well as a programme to be nominated to the annual policy deployment process.

Each team's first activity is to carry out a detailed cost audit to confirm the level of saving achievable and a programme to deliver it. Experience shows that this analysis typically increases the level of benefits identified from that forecast by the loss model. Once the costs deployment process is in place this becomes a two-way process with teams identifying their next improvement target as part of the delivery of their current goal. When integrated with the business-planning process this provides the delivery mechanism aligning accountabilities behind a single change agenda.

Bottom-up focus

At this stage, the typical aims are to reduce the need for intervention during production runs, to achieve 'vertical take off' of quality following

Cost Deployment Summary

Location: Engineering Company
Version: 1.0 11 April 02
Top sheet attached [Y | N]
Scope of system: Component Production
Potential Gain: £560,500

Completed — Tactic Deployed
(4 | 1)
(3 | 2)
Started — Tactic accepted and KPI's in place

No.	Tactic Description	Cost	Forecast Benefit	Resp'	Status
1	Improve OEE to release capacity for new business development. (Overhead reduction)	£10k	£181,500	AF	⊕
2	Standard shift working, (Contributes to 7) to raise lowest shift productivity to average	£5k	£68,000	PM	⊕
3	Standardise planned maintenance and carry out refurbishment to reduce sporadic losses by 25%	£20k	£100,500	AB	⊕
4	Refine/training in core competencies to improve flexibility and reduce avoidable waiting time (Contributes to 2)	£10k	£150,000	CW	⊕
5	Improve bottleneck resource scheduling to reduce avoidable waiting time by 50%	£5k	£54,000	MN	⊕
6	Reduce human intervention during equipment cycle to improve productivity (Contributes to 1)	£15k	£20,000	RL	⊕
7	Improve best practice and technology to halve the quality failures	£5k	£40,000	RL/PM	⊕

FIGURE 7.10 Cost deployment summary example.

changeovers, to stabilise and extend component/tool life to deliver flexibility and outstanding performance. The main targets for the optimisation processes are therefore

- zero contamination (and the need to spend excessive time cleaning) and
- zero defects (and the requirement for technical functions to prevent the manufacture of defects during manufacturing).

Although those of you who have progressed down the Six Sigma route may be concerned about the 'concept of zero', here we use it as a targeting exercise which leads to the progressive creation of optimal operating conditions. In this evolution, some zero goals are easier to achieve than others are.

In a plant making steel pipes, the achievement of zero pipe jams resulted from fitting a mirror so that the furnace could be continuously observed without the discomfort of looking directly into the oven. Within a very short time the conditions which created 'jam ups' were identified and the problem was consigned to the history books. The benefits to the plant and personnel were enormous.

The most important first step was engaging the workforce in the 'zero goal'. The process engineers had already failed to solve the problem with a more technical approach, the idea for the mirror came from a shop floor worker. The movement towards thinking about optimal conditions is a watershed for the Lean TPM organisation. It marks a transition from event driven 'five why' thinking the science of determining optimal conditions. This 'five hows' approach uses proactive versus reactive thinking which demands good technical knowledge and a positive learning environment.

Zero contamination

Zero contamination is an important step to achieving optimised operations. Dirt and dust getting into moving parts will vary component wear rates and make component life difficult to predict. Much of the easy to deal with contamination sources will have been dealt with through the actions described in Chapter 6. Common actions include

- implementing equipment modifications/design standards to minimise scattering of dust and dirt;
- establishing maintenance standards for reassembly of ducting/dust control mechanisms following maintenance and
- working methods which avoid the use of airlines to blow out contamination.

Making significant progress towards zero defects will generally mean raising standards and introducing a new generation of countermeasures at the machine and with the operations teams. In some industries, the benefits of doing so have more than justified the cost of introducing 'clean room' workplaces to replace previously uncontrolled working environments. Eliminating the need to clean therefore recovers time and productive capacity which can be usefully sold in the form of increased outputs and this is an important element of combining a growth strategy with cost minimisation.

OPTIMISE	TYPICAL PARAMETERS
Variation in Operating Environment	Temperature level, Temperature change Humidity. Shock, Vibration Incoming material variation, Voltage change, Batch to batch variation, Operator intervention
Potential For Human Error	Visual controls, Standards, Activity indicators, Alignment markings, warnings, Andon, Audible controls, Mistake proofing
Equipment/Process Deterioration	Corrosion, Brittleness, Dirt and dust in moving parts, Perishing, lubrication, Alignment, Oxidisation Piece to piece variation, Process to process variation, Tooling storage and retrieval damage, Instrumentation calibration, cleaning processes

FIGURE 7.11 Zero defects and optimisation.

Zero defects

The main driver for the optimisation process is defect reduction. Figure 7.11 sets out the main generic causes and examples of the parameters to be optimised.

Using the five Hows

The process stabilisation activities described in Chapter 5 focuses on problem solving and dealing with sporadic losses. As previously mentioned, this process is characterised by the five why's technique to find out *why* something happened. Optimisation is concerned with problem prevention and chronic losses. Chronic losses have multiple causes that are not always transparent at the time. In fact the causes of chronic losses can often deteriorate further to manifest themselves as sporadic failures at some point in the future. Dealing with chronic losses is characterised by an approach to identify *how* the impact of contributory factors can be reduced or removed. Table 7.1 summarises the steps of the five hows technique.

Table 7.2 provides a checklist of points to consider when defining optimum conditions.

In some cases, the chronic loss cause/effect mechanism can be unclear despite significant analysis. In such cases, it is not possible to clearly identify what is triggering off a defect. Identifying the likely contributory factors, defining and implementing standards to reduce their impact provides a practical progress towards the zero defect goal. In some cases, it is easier to begin from the perspective of addressing suboptimum conditions first before beginning the analysis. For example, Figure 7.12 describes a common problem on a bottling line where caps fed from a vibrating feeder jam causing intermittent supply.

TABLE 7.1 The Five Hows Approach

How	Notes
Define 1. *How* and where the target defect occurs	Identify the characteristic of the defect by defining it in physical terms. e.g. dimensional precision, outer appearance, assembly precision.
Identify 2. *How* the defect can be brought under control	Identify the processes that could contribute to the defect. Understand the function of each and assess the conditions that will minimise causes of variation.
Design 3. *How* to control and 4. *How* to reduce defect levels	Observe/experiment to understand which parameters have the most impact on target defect levels (80% of the control is with 20% of the potential parameters). This often needs to include parameters which have not previously been considered as important e.g. heat or humidity, method variation. Introduce low cost or no cost ideas first. Aim to first stabilise and then reduce the defect levels. Focus on ease of detection and early response.
Refine 5. *How* the controls are applied to make it easy to do right, difficult to do wrong	Refine the process to reduce the need for technical judgement without reducing levels of awareness and understanding.

TABLE 7.2 Considering Optimal Conditions

Factor	Potential Improvement Targets
Operational conditions	Processing conditions, operability, conformance to specification
Assembly precision	Precision of assembled parts, vibration, assembly fixtures, interfaces/linkages/timing
External appearance	Dirt, scratches, rust, pinholes, deformation, discolouration, seizure, uneven wear, cracks
Installation precision	Vibration, level/fit
Function	Operation across complete range, compatibility with other parts, actuating systems, intrinsic reliability
Dimensional precision	Required precision, surface finish, life span
Environment	Dust dirt, heat, cleaning process, pipe work, waste disposal routes
Materials/strength	Specification tolerance, storage condition, ageing/brittleness, corrosion

For the technical and operations teams, it was unclear what had caused each jam and there seemed to be no common pattern of failures. Previous attempts at problem solving had resulted in ad hoc adjustment to feeder vibration settings and a return to the traditional 'fiddling' of machinery to try and reach an acceptable outcome. Following the 'five hows' methodology. The process of brainstorming commenced with asking 'how' five times to reach zero loss performance. Here is what the team found, just one illustration of the many they

BIRDS EYE VIEW OF PROCESS

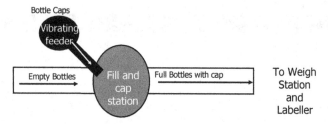

Chronic Loss: intermittent feed of caps to the filler

Frequency: 10 times per shift. Mean time between intervention 4.8 minutes

Impact: when caps are not fed, line stops and needs to be reset.

FIGURE 7.12 A bottling line case.

originated and investigated, and how they ascertained the optimal conditions for equipment operation:

Zero Jams of Bottles during Processing.

1 **How** to define the problem in physical terms (friction of bottle on process conveyor)?

2. **How** to control the level of friction? In some instances, it is necessary to define a precise measure on the phenomena to be achieved. In this case, it was sufficient to look for how friction might be reduced.

3. and 4. **How** to control and reduce friction (countermeasures that could be introduced)? These countermeasures included

   a. alignment of the vibrating feeder and slide to the filler;

   b. routine clean out of the vibrating bowl feeder to reduce the build up of plastic debris on the slide;

   c. cleaning of the slide once per shift; and

   d. setting a maximum level for loading of the vibrating bowl feeder.

The outcome was that, through the combined team effort of applying operator knowledge and technical skills, the countermeasures reduced the jams to zero. The team later found another source of problems (plastic debris eroded from the bottom of the sacks which the caps were delivered in that contaminated the conveyor lines) and adopted the same approach to optimising conditions. Neither of these solutions was expensive and was insignificant in comparison to the benefits delivered. Finally, the teams 'closed out' the problem solving by documenting the approach as a case study.

5. **How** to make it easy to do right and difficult to do wrong?

   a. Visual indicators on the hopper levels and next to the cap storage area. Single point lessons adding the clean out of the bowl weekly and the slide (daily), painted match lines used to highlight bowl/feeder chute alignment.

Having learnt the link between defect prevention and optimum conditions, the team was able to transfer that lesson to other 'problem areas' with great effect and almost zero cost.

Tools to support optimisation

Although the five hows requires a different outlook in terms of problem defi-nition, the asset improvement plan and value mapping tools set out in earlier chapters are still applicable to the process of understanding and resolving opti-misation issues. The major difference in approach, during the optimisation of equipment, lies in the selection of improvement priorities. These are therefore driven by a clear understanding of how the customer makes decisions and what the customer values. The ultimate goal of this stage of the journey is to deliver market/customer leading capability. To support this, a more detailed voice of the customer (VOC) analysis is used to provide the context for product and ser-vice development. This included information which will provide an insight into many more issues that customers take into account when buying products. The categorisation of features is as follows:

- **Qualifiers features** which the customer does not discuss but expects to be offered (these are the most likely causes of customer complaints). Such fea-tures include the implicit understanding that the product is safe and will be of good quality.
- **Winner features** which guide customers' buying decisions and these are features they value and are prepared to pay (more) for. These features are used to compare between competitor manufacturers and are a means of dif-ferentiating your product in terms of better performance than the others.
- **Excitement features** that reinforce customer loyalty by providing hidden value. Such features could include extended warranty for products or the ability to recycle the packaging used.

The analysis also provides guidance on unmet customer needs. To quote Michael Dell 'One of the magic abilities of any great product company is to understand technologies and customer requirements and come up with the per-fect combination to solve a problem. Customer's sometimes do not realise that they have a problem to solve. The customer is not likely to come and say they need a new metallic compound used in the construction of their notebook com-puter but they may tell us they need a computer that is really light and rugged. Where some companies fall down is that they get enamoured with the idea of inventing things and sometimes what they invent is not what people need'

VOC profiling also identifies the sources of competitive drift/advantage that could be exploited. It also provides a measure of current customer relation-ships and how to improve them and a structured template for comparison of the current/planned product and service portfolio. Finally it helps to predict and protect the 'value' offered by the business to its customers.

To illustrate the VOC categories – A tablecloth in a high-class restaurant is an example of a qualifier feature. It is expected and would only be noticed if it was not there. Satellite navigation in cars is an excitement feature and is typically something that would not be the reason for buying the car but would be a welcome feature. Excitement features have a habit of becoming winner or qualifier features over time. At one stage a remote control TV was an excitement feature, now it is very much a qualifier feature. Winner features are those which the customer uses to distinguish between options. In theory, these are the features which the customer would be prepared to pay more for. Higher top speed, wider screen, haute cuisine. It is interesting and profitable to use this form of analysis with teams and to identify the control items on the equipment that must be carefully managed to ensure that qualifiers continue to be met as customer expectations increase. The same goes for winners.

Finding out how customers weight each feature (prioritisation by the customer) is achieved by sets of paired questions. For example, if a customer is asked

- A. how would you feel if the feature is present;
- B. how would you feel if the feature is absent.

If his answer to A is not bothered and his answer to B is unhappy, this is a qualifier feature like the tablecloth in the restaurant. If his answer to A is good and B is not bothered, this is an excitement feature like the satellite navigation. A typical VOC exercise would be carried out with each main customer category to identify

- where low scores on basic features indicates a risk to competitiveness;
- where the potential performance features indicate the opportunity for additional value and
- where 'excitement features' suggest potential unmet needs to use as the focus for innovation generation processes/new product development.

    These features can then be analysed to identify
- how to define the feature in operational/technical terms;
- the link between the feature and the process used to generate it;
- the relevant process variable/parts (control points) and
- control points for those process/part characteristics.

The product/service X matrix (see Figure 7.13 below) is similar to that of the policy deployment x chart and provides a means of representing and understanding relationships. The example below shows an assessment of the interaction of processes which impact the delivery of the key risk/opportunity features from a VOC study. This provides the first step towards understanding what options are available to optimise process parameters. The matrix below also illustrates how the investigation process has been split across multiple functions to facilitate innovation and opportunities for improvement. Although a lead function is assigned to each control area, implementation of improvements will be through the multidisciplined teams starting with the implementation of low-cost or no-cost improvements, with further real-time investigation to identify and justify technical improvements.

The 'X matrix' is therefore a means of 'joining up thinking' and grounding the idea of customer value during quality optimisation processes. It tells you where you need to control the process, what regulator shows the performance of the machine and how to ensure that this variable never changes. This is an important step and breaks an old problem that has affected many Western businesses. This issue was caused when, in the old days, continuous improvement groups were established and they cleared away a lot of low-hanging issues for operations

The X matrix — upper quadrant (Action / Feature)

Proc	QA	Maint	Ops	Log	Purch	Feature		Mix 28%	Shaper 15%	Dry 20%	Kiln 23%	Pack 15%	
						Cosmetic Appearance	Basic	2	2	3		3	
						Predictable Life	Perf	3	1	2	3	1	
						Long Life	Perf	3	1	2	3	1	
						Strength	Perf	3	2	1	3	1	
12%	17%	22%	17%	17%	17%	Total		11	6	8	9	6	40

The X matrix — lower quadrant (Parameter/Part)

Centre diagonal labels: Feature · Parameter/Part · Control Point · Process

Temperature	Mixing time	Tool wear	Cleaning	Restocking	Suppliers	Parameter/Part	Mix	Shaper	Dry	Kiln	Pack	Total	
				5	5	Raw Material Quality	2	1	1	1		5	10%
	2	2	3	2	1	Dust Control	2	1		2	2	7	14%
	5	3		1	1	Cutting		3	1	1	2	7	14%
2		2	3	1	2	Start Up Routine	3	1	2	3	2	11	22%
2	3	3		1	1	Steady State Routine	3	2	5	3	3	16	32%
3		3	4			Close Down Routine		2	1	1		4	8%
7	10	13	10	10	10	Total	10	10	10	10	10	50	

FIGURE 7.13 The X matrix.

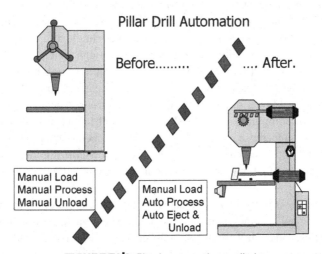

Pillar Drill Automation

Before.........   .... After.

Manual Load
Manual Process
Manual Unload

Manual Load
Auto Process
Auto Eject &
Unload

FIGURE 7.14 Simple automation applied.

teams but, over time, the lack of technical skills of the operations teams meant that they could describe problems but did not have the skills to solve them (these were in the heads of technicians and engineers). Soon, continuous improvement became discredited as operators complained and the technical services, which did not hold these issues as high priority, eventually got around to improving the problem. However, by this time, with no real problems that could be solved, the continuous improvement teams either turned into social activities (legitimate skiving) or the programmes collapsed and were discredited. The power of cross-functional teams is therefore enormous and the targeting of improvements must therefore come from the customer and the analyses undertaken by the firm.

TABLE 7.3 Forms of Simple Automation

Level of Automation		Load Machine	Machine Cycle	Unload Machine	Transfer Part
1			Manual		
2		Manual	Automatic	Manual	
3		Manual	Automatic		Manual
4		Manual	Automatic		
5			Fully Automatic		

Source: Toyota Motor Corporation.

Low-cost automation

Once stable operation has been achieved, improvement activities can focus on the use of low-cost automation. Generally automation which merely replaces human activity will take many years to payback. There are generally opportunities to apply low-cost automation to reduce the need for essential nonvalue-added activities such as inspection, to coordinate inter-site material movement and simplify start-up processes. Such programmes should aim to progressively increase in-house capability using a 'learning through doing' approach. Not only does this support the horizontal empowerment approach but it reinforces the important lesson that simplification must precede automation (Table 7.3 sets out five potential levels of automation). Progress through each level adds to the complexity of equipment and the risk of equipment failure. It is important that this is met by an increase in capability. Once stable operation has been achieved, improvement activities can focus on the use of low-cost automation. Generally automation which merely replaces human activity will take many years to payback.

The loading activity usually contains the highest level of complexity. This is where positional accuracy is most critical and where variation in materials can result in jams, and quality problems at upstream processes. Increasingly, this level of complexity is seen as a step too far by many 'world-class' companies (including Toyota) and full automation is only ever possible when the last issue for optimisation teams has been solved (the point at which the lights on the business can be turned off and for the machines to work without any form of interruption). And this brings us to a further capability that is needed to achieve the optimised state and that is optimisation of new assets to be installed at the factory – this pillar is known as early equipment management (EEM) or just EM for short.

7.5 EM APPROACH TO CAPITAL PROJECTS

New equipment should be capable of achieving normal conditions from production day one (No breakdowns and stable running without unplanned intervention). They should be capable of low life-cycle costs (LCCs) and able to deliver

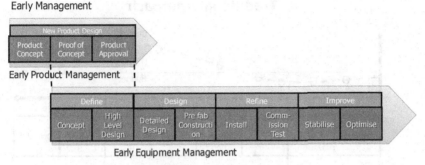

Early Management

FIGURE 7.15 Early management.

current and future customer needs for products and services. Unfortunately most managers and engineers have had first-hand experience of capital projects that failed to live up to expectation when introduced and needed significant attention during routine operation. The excess capital costs of these troublesome assets can be huge. Opportunity costs are high too. One organisation we supported estimated that improvements in capital project delivery was enough to recoup the original capital investment during the first year.

EEM was developed as part of the TPM tool set to meet these challenges by applying the same improvement logic used for manufacturing to the delivery of investment projects. The aim being to improve the effectiveness of the process (and systematically deliver a process) which is easy to do right and difficult to do wrong by design. As TPM evolved so did the EEM tools set to include the challenges of product design and design for manufacture. The combination of EPM (Early Product Management) and EEM is known as EM (Early Management).

EEM and EPM are linked by their stage gate process as shown in Figure 7.15. Often the EPM precedes an EEM project, particularly if the new product or service requires investment in new equipment. Together, as EM they provide a focussed improvement route map to reduce time to market and deliver assets which are flexible to market development activities and rapid demand growth.

Frequently it is the justification process for new product or services that triggers the call for investment. Without due consideration of equipment needs, that can mean that the decision to invest in new products is flawed from the start. Although EM concerns capital projects, it is supported by systems to capture the lessons learnt from taming today's product and technology issues so that they are eliminated from future customer offerings. It is also supported by improved investment planning so that opportunities released by optimising current technology can be fed back into enhanced products and services.

The Toyoda EEM experience: case study

Figures 7.16 and 7.17 set out the differences between the traditional project approach and that of EM. This is based on the experiences of Toyota. The

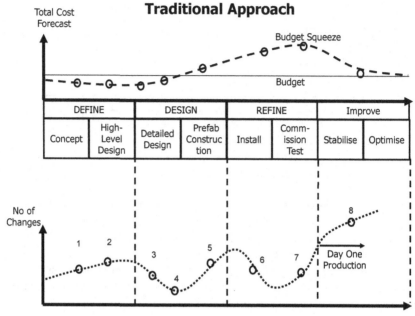

FIGURE 7.16 The traditional approach to new assets.

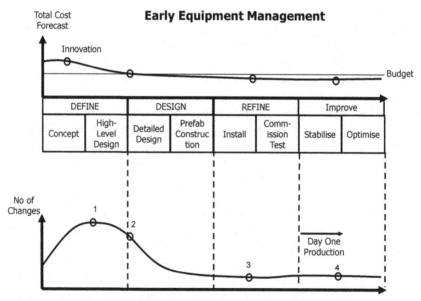

FIGURE 7.17 The EEM approach to new assets.

figures compare similar projects pre- and post-EEM. Figure 7.16 shows the traditional route. At points 1 to 2 of the change curve at the bottom of Figure 7.16, input is sought on the design but once funding has been approved at the high-level design stage gate, changes are resisted for fear of scope creep. It may be that at this stage the vendor is on a fixed fee to deliver the project so they will seek additional costs for any changes made. In addition at this stage, there is pressure to get the design signed off so that orders for long lead time items can be placed. The failure to trap latent design weaknesses at this stage means that they will occur during the construction and installation/commissioning stages. Unexpected problems this late into the project are very expensive to resolve. As total project costs estimates increase (Figure 7.16 upper graph) to bring the project in on cost, budgets are squeezed. This can mean that spending on areas such as training and spares costs are sacrificed. The outcome is a project which fails to deliver its promise on day 1 and for months if not years afterwards.

Figure 7.17 sets out their experience after introducing EEM. At the start of the project at point 1, the forecast costs are higher than budget but the insight gained as a result of using EEM tools indicates potential to reduce overall capital costs. Confidence about the level of improvement possible through innovation also increases as organisations gain experience of using EEM. From point 2 onwards the search for improvement continues (even though the funding budget has been set for a number of reasons) including the following:

- It is not possible to specify exactly what a new process will look like at the beginning of the project. There needs to be some allowance for a learning curve. Clearly there needs to be control of costs but cost estimates can never be 100% accurate until the detailed design is finalised. Collaborating with vendors at this stage is a powerful way of capturing latent design weaknesses and improving project deliverables. Trying to pass off as yet unknown risks to a vendor is a recipe for higher costs later when those risks turn out to be related to issues outside of the vendor's control.
- EM seeks input on innovations to reduce LCCs and seeking opportunities to gain added value from the process.

The last point is important. The focus of any design is to deliver the required outcome to 'right size' and 'right spec' assets (that will be around for many years). Once the detailed design is complete, looking for additional benefits from the design is a lucrative source of added value. The aim here is to enhance project value within the existing budget rather than to increase the scope of the project spending.

The benefits of EEM

The benefits of EEM are significant. In addition, the gains from adopting EEM principles as part of your capital project programme include lasting gains in terms of

- Development of in-house capability to deliver
  - flawless operation from day 1;
  - low operational LCCs and
  - increased return on investment.
- EEM captures and unlocks tacit knowledge to support
  - cross-project learning;
  - clarity of investment priorities;

- project ownership and
- innovation.
- EEM is an improvement process for projects to
  - speed up project delivery;
  - improve collaboration across functional/company boundaries and
  - improve design, specification and project management systems and processes.

Research into the causes of capital project failures indicate a number of common failings at the early projects steps including

- Missing or badly designed user specifications so that
  - project accountabilities and roles are ill defined;
  - avoidable risks and preventable problems are not identified and dealt with until too late in the process.
- Resources released late or not released
- Insufficient training and support for project stakeholders
  As a result problems occur at later steps including
- conflicting views of what is needed;
- lost opportunities to challenge and optimise design choices;
- critical decisions were delayed or not taken and
- communication between project stakeholders interrupted or lost.

As can be seem from the above, the root causes of problem projects are not generally technical issues but relate to people and how they work together. The root causes of many of these issues can be found in the 'behavioural tics' of framing, optimism and the use unproven rules of thumb introduced in Chapter 1.

EM (equipment and product) counters these tendencies through the use of

- the structured collaboration of commercial, operational and technical teams using a structured stage gate decision process;
- qualitative EEM design standards, checklists and evaluation processes surface latent design weaknesses early and enhance project value and
- formal stage gate reviews at key decision milestones to avoid the transfer of problems to later steps.

EM stage gates are designed to support the following project governance goals during the project life cycle (set out in Figure 7.14):

1. **define** the correct scope of the project and involve the right expertise;
2. **design** the process in a way that provides the insight to tease out latent design weaknesses and enhance project value;
3. **refine** the shape of the new operation, master new competencies to deliver low LCC operation and achieve vertical start-up of the new operation and
4. **improve:** successful delivery of flawless operation achieves process stabilisation. For that reason following day 1 production, the goal of EEM is optimisation.

Delivery of these goals is supported by principles and techniques to avoid the common pitfalls that befall capital project delivery as explained in the paragraphs below.

Define stage gates

The root causes of later problems can be due to weaknesses in the way investment priorities are set and project scopes defined. For example, a company at the start of its EM journey decided to invest in upgrading equipment to bring in-house a product which currently was packed by an outside company. A cost hike from the outside packer had recently focussed attention on the need to curtail this increased cost.

As part of Step1, the EEM core team identified that this was not the best use of the investment and that if the funds allocated were spent elsewhere it would produce greater benefits. Normally, the original investment request would have been carried out without question but following this review a much larger project with a faster payback was identified.

Based on this experience, investment projects briefs were then set out in terms of the desired outcome and prioritised targets. The EEM core team role was to
- translate targets into metrics;
- define problems to overcome to deliver those targets;
- set out scenarios to consider and
- evaluate those scenarios and develop a preferred approach.

At this stage of a project there is often little qualitative data to support the evaluation of scenarios. There is a need to codify tacit knowledge to make and objective choice between options.

EM supports this by the use of six qualitative design targets, each of which has a significant impact on LCCs but can be difficult to measure even with hands on experience. These six target areas are set out in Figure 7.18. This also shows the 1–5 scale used to calibrate assessments. (1 being low, 3 being acceptable and 5 being optimum). Where necessary, following this qualitative assessment, further analysis can be carried out to confirm the evaluation basis, refine the preferred concept and support the submission of the application for capital project funding.

Design stage gates

Following funding approval of the preferred option and selection of a vendor, the priorities for EEM Steps 3 and 4 are to tease out latent design weaknesses and enhance project value. Common mistakes here are trying to pass off risk to the vendor without a clear understanding of what the risks are or whether the vendor is able to manage them.

Experience shows that there is more to gain from collaborating with vendors to develop better designs. Even where the options to modify technical design are limited, the understanding gained about changes to operations, workflow and skills needed etc. make this an important activity on the road to flawless operation.

Our analysis shows that the difference between an 'average' and 'simplified' design can be worth a 30% reduction in capital costs. The same analysis shows that 'complex' designs can increase capital costs by more than double when

	Definition	3. Acceptable	5. Optimum
Safety and Environmental	Function is intrinsically safe, low risk, fail safe operation able to easily meet future statutory and environmental limits	Little non standard work / Moving parts guarded, few projections / Meets SHE and fire regulations / Easy escape routes and good ergonomics	Foolproof/failsafe operation / High level of resource recycling / Uses sustainable resources
Reliability	Function is immune to deterioration requiring little no intervention to secure consistent quality	Low failure rate / Low idling and minor stops / Low quality defect rate / Flexible to technology risks / Good static and dynamic precision	High MTBI / Stable machine cycle time / Easy to measure / Flexible to material variability
Operability	Process is easy to start up, change over and sustain 'normal conditions'. Rapid close down, cleaning and routine asset care task completion	Simple set up and adjustment mechanisms / Quick replace tools / Simple process control / Auto load and feeder to fed processing	One touch operation for height, position, number colour etc. / Flexible to volume risk / Flexible to labour skill levels
Maintainability	Deterioration is easily measured and corrected. Routine maintenance tasks are easy to perform and carried out by internal personnel	Easy failure/detection/repair / Off the shelf/common spares used / Long MTBF, Short MTTR / Easy to inspect and repair	Easily overhauled / Self-correcting/auto adjust / Inbuilt problem diagnostic / Predictable component life / Fit and forget components
Customer Value	Process is able to meet current and likely future customer QCD features and demand variability. Provides a platform for incremental product improvement .	Easy order cycle completion / Maximum control of basic and performance product features / Flexible to product range needs	Capacity for future demand / Robust supply chain / Simple logistics/forecasting needs / Flexible to potential market shifts
Life-Cycle Cost	Process has clearly defined cost and value drivers to support Life-cycle cost reduction, enhance project value and maximise return on capital invested	Clarity of current capital and operational cost drivers and process added value features / Potential for value engineering gain / Resource economy	High level of resource recycling / Flexible to financial risks (e.g. vendor) / Easily scalable to 400% or to 25% / Access to high added value markets

FIGURE 7.18 EEM benchmarks.

compared to the 'average' design. That alone is enough reason to take care at the design review stage. However, capital costs are normally a small percentage of the total LCCs and achieving flawless operation from day one can be worth as much again.

EEM design collaboration involves a two-part review process:

- meeting 1 involves assessment of an outline design against EEM standards to confirm strengths and priorities. Use 'day in the life of' approach and layouts to make clear operational realities and outline ways of working and
- meeting 2 involves the vendor explaining in detail how the more detailed design meets those standards.

That way the vendor can show that they really do know what is needed to achieve flawless operation from day 1.

Refine stage gates

After completion of the detailed design process, human error is the biggest risk to the capital project. EEM Steps 5 and 6 support the evolution of the new operation and management of the glide path to flawless operation. A common mistake here is relying on vendor training alone to develop in-house competence. Competence is achieved over time and needs reinforcement. Training alone cannot achieve this. In addition, when introducing new equipment, this is the ideal time to engage representatives across shifts to agree a common approach.

Within an EEM project, the evolution of ways of working is managed as a core part of the specification process at each step. That is the only way to be sure that Morse principles (maintainability, operability, reliability, safety and environment) are hard wired into the design. To further refine ways of working problem prevention techniques are applied to provisional ways of working so that they can be easy to do right, difficult to do wrong, simple to learn. This incorporates the PDCA (plan, do, check, act) approach by testing ways of working as part of the commissioning process.

Implementing early management

Like EM itself, successful implementation programmes adopt a stepwise approach:

1. **Define:** Review current capital project processes and identify strengths and priorities for action.
2. **Design:** Use a pilot project(s) to improve the current processes, define EEM standards and deal with gaps in related business processes.
3. **Refine:** Train managers, design engineers, project managers and CI facilitators in lessons learnt from the pilot.
4. **Improve:** Capture lessons learnt from each project and use to update design standards. Reinforce EEM policy by training of each new project delivery teams including key stakeholders.

Design to life-cycle costs

Another important differences between traditional capital projects and EM is the use by EM of LCCs rather than just capital costs. LCC analysis combines capital and operational costs to set the context for decisions at all project stages. The process of design to LCCs sets out a unifying challenge that broadens the frame of reference of project personnel from delivering the capital project to delivering the project business case. In some instances, the operational costs can swamp capital cost as shown in the pie chart at Figure 7.19.

Here the gain made by reducing hidden losses were 7 times the original capital costs. This was only achieved by revealing this profile to the project team. It completely changed their outlook and motivation. It should be noted that the life-cycle models in this context are much less complicated that the models developed by financial analysts to forecast life cycle. Here we are simply trying to assess the main contributors to LCC so that when choices are made, that is informed by the impact on LCC. EM typically uses simple models that show the difference between options or against the current base case rather than a precise prediction of LCC. Typically we use simple nondiscounted 5-year forecasts of operating costs based on current budgets. This avoids the need to build a model from a zero base and as long as we compare options against the same base, the decision will be sound. Naturally where options reveal similar LCC flows, it may be necessary to increase the sophistication of the model but at the early stages of the project, when you do not know what you do not know, it can be sufficient to compare options against the six EEM target areas using the criteria set out in Figure 7.18 above. Later on as detail is added and more information becomes available the LCC model is refined adding detail to support the decisions required at each step.

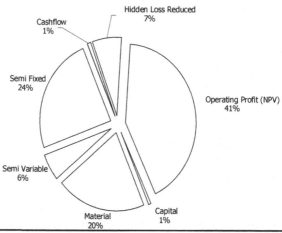

Life-Cycle Cost Profile Example

In this example, over a 5-year period, capital costs are a small part of Life-Cycle Costs

FIGURE 7.19 Life-cycle cost profile (example).

7.6 LEAN TPM CAPABILITY DEVELOPMENT

Achieving success with the delivery of optimum conditions needs a combination of technical and people skills/behaviours as set out in Figure 7.20 below.

These conditions are best developed through initial foundation training, such as that set out at Figure 7.21, quickly followed by supported application during practice projects. To select those projects, it is important to revisit the value stream map of the key products manufactured at the factory. These maps clearly show where best to invest these skills in project work. The ability to 'see' the impact of improvement activities within the value stream helps determine the commercial benefits of the exercise – a rare phenomenon with most improvement activities that do not yield a commercial benefit to the firm. Furthermore, the development of the basic Lean TPM skills sets and then advanced techniques for squeezing the last remaining elements of waste and chronic loss helps to break through the traditional glass ceilings that have prevented the meaningful involvement of machine operators with process improvement activity. These skills, especially when new products have reached a stable state of order volumes, will be used many times during the service of the individual. Also by creating a critical mass of these skills, the entire class of operators will be capable of sustaining their own improvement efforts. By consequence, this means operators will be able to move to a new level of problem solving. At this stage these employees can identify (using cause and effect charts) problems with the machines they operate and can select the right tools to analyse these problems in greater detail and then to solve or reduce these barriers to improvement.

	Techniques	Behaviours

Techniques
- Standard Setting
- Problem Solving Tools
- Information and Analysis
- Cause/Effect Mechanisms
- Value Stream Priorities
- Goal Clarity
- Team Design/Profiling
- Functional Skills
- Recognition Processes
- Action/Review Processes

Behaviours
- Personal Values
- Identity
- Principles
- Collaboration Skills
- Supportiveness
- Openness
- Systems/Process Integration
- Learning Process
- Interdependency
- Influence

FIGURE 7.20 Technical peak performances.

	TPM SKILLS	Lean Skills	Leadership/Learning Skills
1	Assessing Value Generating Processes, Cause/Effect Mechanisms, Gap Analysis, Strategy Development and Standards Setting	Defining value from the customers perspective, macro mapping current and future state	Understanding leadership, agreeing future destiny, defining new patterns of behaviour, redefining the rules of the game.
2	Goal Clarity, Prioritisation, Accountabilities, Formalise Practices and Problem Solving Tools	Using process mapping to understand where value is added	Setting new priorities and focus, raising standards building empowering relationships, mapping value and action plans
3	Problem Prevention Tools, Cross Functional Processes, Recognition, Action/Review Process,	Improving process flow	Establishing high performance team values and identity to improve new standards, developing trust and self correcting team processes, conditioning strategies.
4	Managing Changes in Working Practices to Sustain Zero Breakdowns and Address Optimisation issues	Establishing new ways of working to address the seven wastes and hidden losses	Managing results not activities, performance review process, the meaning behind performance, using information to energise and maximise productivity

FIGURE 7.21 Skills for lean TPM mastery.

Perfection: the extended value stream

Activities to deliver milestone 4: Strive for Zero (Perfection) aim to complete the process of changing the competitive landscape to lead the customer agenda for chosen products and services. This builds on the capabilities developed in milestone 3.

Exit criteria includes
- delivering industry leading standards of delivery of customer value;
- established strategy to disrupt current competitive landscape and control the rate of change for other organisations in the same sectors and market segments and
- world-class standards of product innovation and customisation strategies.

An important variable, which must be controlled in order to achieve a high level of material flow, is that of supplier quality assurance. So far we have discussed employees and machinery as inputs to high performance but, for most companies, supplier management routines are poorly developed. But rarely, at any company has it been written that production operators, quality assurance personnel or maintenance engineers should remain in the factory. Indeed this unwritten constraint in management thinking will slow the achievement of perfection if it remains unquestioned. When all is said and done, these inputs represent materials that happen to be bought in rather than made. As such, suppliers represent 'associated' manufacturing facilities and should be regarded as collaborators in the process of achieving high performance. All too often suppliers are regarded as elements of the production system that should only be treated as 'enemies', to be beaten up for cost reductions and never to have information shared across the two companies so that this information is not used to force higher prices. This view is somewhat outdated (whilst remaining common) by companies engaging in lean practices and the lean approach.

All major Western manufacturers face a growing deficit in the training and quality of new engineers, quality assurance personnel and other employees. This is concerning and potentially creates a vacuum in the supply of new talent to drive the process of 'perfection'. In the future, this means, either manufacturers will have to buy in consulting engineers to help improve processes (even though these individuals do not operate the equipment) or share this talent with customers and suppliers in a form of cross-transfer of people to assist supply chain material flow improvements. This latter activity affects not only the 'door to door' overall equipment effectiveness measure but the supply chain flow performance. For most businesses, the materials element of product cost is significant (usually between 60% and 80%). When a total costs of supply approach is used the cost of poor quality, needing to inspect supplied products, the costs of failures at the customers internal value stream and other costs means enlightened business managers will share engineering talent at the very least.

The approach to co-managing the value stream is not new and has been associated with not simply process optimisation but a source of competitive advantage. Here too there are many falsehoods:

1. suppliers are less sophisticated than their customers;
2. suppliers should be kept at 'arms length' and prevented from involvement in improvement activities;
3. suppliers are disinterested in joint improvements with their customers;
4. multiple sources of supply should be maintained to increase customer power when engaging in price negotiations;
5. suppliers should be rotated regularly to avoid complacency in the trading relationship;
6. short contracts should be offered to suppliers to ensure the ability to frequently exert power in price negotiations and
7. suppliers will use information to increase prices.

The most common approach to integrating suppliers by the major Japanese manufacturers is that of the 'supplier association'. The supplier association is a forum of the key suppliers to the customer and is hosted by the customer organisation in the form of a series of meetings each year used for both sharing and co-ordination of supply chain improvement efforts and also for supplier development. The forum is an important lean mechanism and is therefore used to transfer knowledge between supply chain partners. This knowledge transfer is legitimate because the customer is dependent upon its supply base and in a lean enterprise suppliers represent an important source of innovation and have a direct bearing on the competitive position of the customer organisation.

Putting it simply there are few better mechanisms than the supplier association in targeting and improving the aggregate performance of productive materials suppliers in the value stream. In parallel, it would be prudent to establish a maintenance association whereby all the chief engineers of important suppliers (rather than commodity suppliers or distributors) gathered together to formally transfer best practices. Forming such a group is important as it leads to better decision-making, a focus on the quality, delivery and cost of supplies and the integration of suppliers with the main challenges of the customer organisation. Too many Western companies 'shy away' from direct supplier involvement but it is important to understand that, when you purchase products, the optimisation of your production system is dependent upon these suppliers. Under the condition of dependency it is logical to co-operate with suppliers and to investigate the best ways of receiving products and getting them to flow through the production system. Surprising still is the willingness of suppliers to help customers improve and therefore to grow even after years of low supplier involvement and the 'power games' associated with trying to 'get the best price'. In the process of optimisation, supplier involvement is important especially when the goal of the entire supply chain is to optimise material flow and compress the cash-to-cash cycle. It is no wonder that these groups, external quality circles if you like, have been used extensively by 'world-class' Japanese manufacturers like Toyota and have been successfully transferred to their Western operations – the benefit of customer and supplier alike.

7.7 CHAPTER SUMMARY

Building on the success of achieving a stable operation requires a change of outlook at all levels, in particular the need to move away from action stimulated by operational fire fighting to that stimulated by a passion for improvement. This process will be frustrating and ultimately unsuccessful unless it is led by a desire to transform the business. As such, change is as much a matter of cultural design as it is technical improvement.

Whilst the technology employed by the factory may be the most sophisticated in the world, it is ultimately the workforce who determines the efficiency of the factory. For employees, it is important to take the view that the value of the typical operator must be continuously enhanced so that a lifetime of improvement contribution can be extracted and the quality of working life improved for the individual. Modern manufacturing has less and less room for 'just a pair of hands' and increasingly demands more and more technical/interpersonal skills.

The value of employees – the most important source of value to a business and although the mantra that 'employees are our number one asset' rolls easily off the tongue, the reality in most factories is that employees remain under conditions that have little changed since the 1920s. The elimination of chronic losses and the optimisation of processes (the other assets) are determined by the employees. In fact, factory efficiency is the result of the employees. Even the most up-to-date machinery will have a reduced efficiency if employees do not know how these assets work and what signs of abnormality can be used to detect failings and take the appropriate countermeasures. Not just that, no business stands a chance of optimising what it does if the skills of the employees are not improved and constantly improved to allow waste to be identified and reduced. To some, this may seem a bit of a waste in terms of giving employees these highly sensitive diagnostic skills but there again these are also the managers who would have believed they had achieved a 'lean business' well before this stage. It can be no surprise that Toyota – the benchmark of all lean businesses continues to promote 'good products and good people' as the basis for its almost 60 years of improvements.

To make this a reality, the top-down leadership challenge is to spring the strategy trap. The bottom-up leadership challenge is to secure delivery of horizontal empowerment. With these dimensions in place managing the technical optimisation process can then deliver meaningful development of human potential and competitive advantage.

Process optimisation is a never-ending process within the lean enterprise. This is supported by the TPM methodology which combines with TQM (Total Quality Management) to add a reliability of material flow that the lean production system (and TQM) cannot achieve by themselves. High quality and low buffers cannot operate indefinitely without TPM. Lean TPM therefore represents a natural extension of these activities that ensures the minimal level of system buffering (to allow free-flowing materials) is supported by a programme of change which focuses on the removal of the inherent failings of the productive asset base. Any future state production system is therefore ransomed by poor or erratic machine performance (especially that of the bottleneck operation). Further, early attempts to optimise the production system will have to contend with existing technology and the TPM optimisation methodology allows both a passage through a human-centred 'glass ceiling' of skills plus a structured means of engineering-out the failings of current and future generations of asset.

Process optimisation therefore targets zero losses and the 'zero-loss' environment is the end-goal of the lean production system within which there are no losses to quality, delivery reliability is assured and costs are at a minimum. These latter features are all, at minimum, 'order qualification' processes that yield customer satisfaction. As such, lean represents a 'why?' businesses are engaging in process optimisation whereas TPM is one of the major 'hows' that the business must engage to create these outputs from the value stream.

The lean approach stresses 'value' as a core business objective. The value of machinery is measured in zero lost time, the elimination of chronic losses and the production of perfect quality products from process capable machinery. However this is not the entire solution to the achievement of 'world-class' performance. The value in closer links with suppliers must be recognised. Buying from cheap sources or businesses that lack the skills to eliminate their chronic production losses will affect the performance of production lines. 'Beating up' these supply sources is hardly a recipe for focusing the minds of suppliers on eliminating waste and enhancing the value they provide. Instead it will result in illogical cost reductions and often the presentation of a price reduction to 'make up for' the lack of improvement activity. A good starting point for suppliers is to show them the waste in their systems by value stream mapping them, using their data to highlight their issues. Furthermore, just because a supplier has an implemented quality management system you should not be fooled into thinking they have the improvement processes nor capability to provide you with high levels of customer service, defect free supplies or a year-on-year set of improvements.

REFERENCE

Hammer, M. (1996). *Beyond re-engineering: How the process-centered organization is changing our work and our lives*. New York: Harper Collins Business.

Chapter | Eight

Moving beyond the Factory

8.1 INTRODUCTION

It often comes as a surprise to manufacturing businesses that the supply chain in which they operate is actually a major source of improvement and innovation. The skills of Lean TPM and the learning that has been generated this far is more than capable of being transferred to suppliers and distribution partners. And why would not you transfer these capabilities? Some people would say that it is the suppliers' responsibility to do their own improvement and some would say that suppliers will just use the methods and try to profit from them. These are natural perspectives but they miss one vital point – at the end of the day we are, like it or not, dependent on the supplier. Without the part they supply we cannot build a finished product. Take car wing mirrors for example: no wing mirror means an incomplete car and rework – costly rework – as well as a delay in cash flow. Ironically, many businesses that go bankrupt are profitable but they have not managed cash flow and run out of money. For those people who struggle with this – you must get paid for the goods you make and sell – you can then buy more materials and start the cycle again. Slow this cycle and you can quickly run out of money or end up spending lots of money without banking the money that should be coming in. What we have learned thus far is Lean TPM shortens the making to cash cycle.

OK so let us go back to the question of whether we should develop suppliers with the same methods and approaches. There are many reasons why the answer is – You Have No Option! If a supplier goes bankrupt you do not get to sell products and therefore you do not get paid so you cannot bank the profit.

Imagine a supply chain that has the same stability and good overall equipment effectiveness (OEE) performance to support its customers as has been mastered by your Lean TPM journey. Now you really can get inventory and stock on wheels! That means keeping materials moving in very small batches. It means personalising products and engaging in a full supply chain redesign. Extended Lean TPM is a major step forward and further reduces noise in the entire supply chain – almost all of the product recalls of automotive vehicles could have been prevented if an extended Lean TPM approach had been undertaken. And you would not believe how many product recalls happen in this product category – the second biggest

197

amount of money you will part with after that of buying your own house so you would have thought the industry would get it right.

8.2 WHY ENGAGE THE SUPPLY CHAIN?

Traditionally, suppliers were held at arm's length from a customer – the logic here was to keep the supplier away from any information or privileged position that may allow the supplier to raise its prices and exploit the customer. This was known as a 'zero sum' game – academic speak for a game of negotiation that meant one business would win and the other would lose. More often than not – the competition amongst supplier businesses would place the relationship power in the hands of the customer and the relationship was adversarial. Basically, at all costs the customer business would use power and withheld information to push supplier prices down. In addition, the customer organisation would design the product to be made by the supplier so the only outcome was the supplier price would go down but their costs would stay the same and inevitably the supplier would be giving away its profit. As profits went down, there was less money to bank, less money to invest in new technologies, etc. The result was inevitably the supplier would trade off a loss of profits by substituting it with a lower level of quality control. In fact suppliers would engage in all the behaviours that prevented the just in time system from being designed and operated.

More enlightened businesses saw their suppliers as sources of innovation and partners in the order delivery and customer satisfaction process. Belcher (1997) also saw the specialist knowledge of the supplier as a means of innovating and realised that a supplier who could design their own product could take costs out of it without forsaking profit. These practices included allowing suppliers to design parts using less materials or less labour to make the product or even an alternative material. The automotive world is a good example of suppliers getting engaged to reduce the weight of products so that the vehicle could achieve a higher level of miles per gallon (and improve the environmental performance of the vehicle).

The evolution of supply chains from transactional, hands-off relationships towards collaborative relationships has been shown to produce gains both in terms of cost and service level performance as shown in Figure 8.1. As mentioned in Chapter 7, Optimisation, collaborating with strategic suppliers and customers is an important growth accelerator through shared new product and service development.

Now there are a few additional issues that must be explored before the process of extended Lean TPM can be really grasped. These are that the purchasers employed by customer organisations are few in number – often less than 1% of the entire company workforce but in product design costs it is often the case that bought in materials and parts can account for around 80% of the final product cost. So there are few skills available in staff terms in the customer business to work with suppliers and find ways of transforming the performance of the supply chain or reduce design costs. So just as operations staff and maintenance staff have worked on a relationship to create the 'zero loss' production system

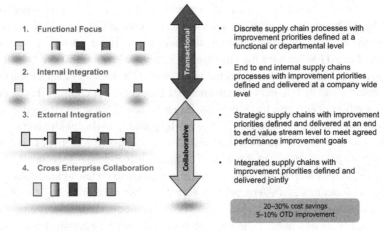

FIGURE 8.1 Evolutionary steps of supply chain relationships.

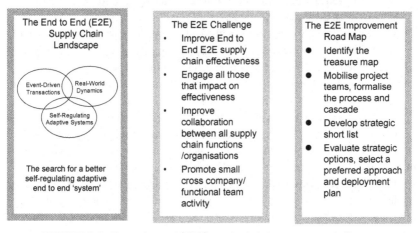

FIGURE 8.2 The end to end (E2E) supply chain improvement challenge.

then it is likely that these same staff will cross the traditional organisational boundaries of departmental-focused businesses. The migration of operations and maintenance staff to work with suppliers is completely common sense when you regard the supplier as an integral part of the fulfilment of customer orders by providing materials when they are needed, in the quality required, in the quantity needed and to the place they will be used. So for a business that has worked with Lean TPM it is quite natural to see staff leave (on secondment) or to support change in a supplier business.

In Figure 8.2, the supply chain landscape box sets out how the essential role of a robust supply chain is to buffer events such as customer demands against real-world dynamics by providing some form of self-regulating adaptive buffer. Traditionally this buffer would have been time or inventory or a combination of the two. As explained in Figure 8.1, collaboration is a better basis for buffering logistics against real-world dynamics.

The second box sets out the challenge and how Lean TPM approach meets that challenge by engaging all those who impact on End to End (E2E) effectiveness with the goal of removing hidden losses and wastes. These wastes are described in the Lean TPM treasure map in Chapter 2. You will realise that the Lean TPM change mandate evolves towards the value chain E2E losses as the challenges of equipment and door to door effectiveness are brought under control. This is really an iterative rather than a linear process. It may be necessary to tackle supply chain issues early in the programme. It may not be possible to deal with supply chain issues until resources have been released by the stabilisation activities. This will vary from programme to programme. The final box in Figure 8.2 sets out the Lean TPM E2E road map progressing from the treasure map review to the deployment of cross-functional, cross-company projects to deliver specific strategic goals.

In summary the principles of Lean TPM E2E supply chain improvement are:

- Supply chains need to be capable of dealing with uncertainty and messy problems to avoid:
    - failed transactions which result in increased system load;
    - extended customer service lead times and rework; and
    - loss of customer cooperation and organisational morale (systems overload).
- System failures increase system load and complexity e.g. reconciliation of data, failed service/complaints or mechanical breakdown.
- The highest performing supply chains are those that optimise 'E2E' system effectiveness rather than individual links in the chain.
- The rate of systems flow is an indication of the 'health' or effectiveness of a system; delays, service failures and rework are symptoms of poor systems health.

High levels of supply chain effectiveness are achieved through continuous improvement rather than a one-off investment in technology or infrastructure.

However, even with available resources (freed from the stable and optimised production process) there still is not enough to provide one member of skilled staff per supplier to act as a mentoring system. So the skills of Lean TPM – some of the most advanced technical knowledge of how to control and master a production system for high performance – must be shared in order to impact on the aggregate performance of all direct suppliers. Now to impact on the average performance of suppliers it is often wise to work on the worst of suppliers because it is these organisations that are holding the entire supply chain back. Not just are they holding the supply chain back but the customer business must cover for the lack of improvement by holding lots of stock – unnecessary stock, a cost to the customer and a loss of space that could be deployed to be used as an area for another production process.

Sometimes suppliers can be more advanced than their customers. Imagine a Lean TPM business that has customers who have yet to go through the process or new customers who could benefit from being involved with this form of improvement activity. These businesses need to be developed too. By showing a customer how to stabilise their assembly or production process so that they could put more money into the bank or speed up their cash flow is a major feature that wins business and differentiates suppliers (Goldratt, 1984).

With these conditions, it can be no surprise that the only real option to companies is to redesign the supply chain as a collective – a team approach if you

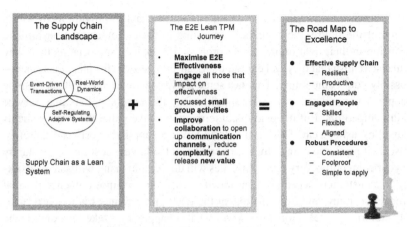

FIGURE 8.3 Lean TPM supply chain improvement.

like. In fact it is exactly a team approach because suppliers are part of the team too – if the company wins new business then the supplier gets more business and more business is cash flow.

The natural outcome of any policy deployment cycle is to create the 'growth challenge' and this means challenges can be deployed to the supply chain. If you want stockless and flow supplies direct to the point of use in the factory, then it is important to develop suppliers in Lean TPM capabilities. So the most cost-effective way of developing the overall performance and capabilities of suppliers with a limited resource is to create a high-performance network using the Lean TPM supply chain improvement approach where engaged people collaborate across the supply chain to develop better ways of working that offer advantage and new value to its members (Figure 8.3).

The network approach is unconventional and, at first, quite daunting. Putting a lot of suppliers in a room who all have a history of supplying the customer raises concerns that one or two suppliers will be disruptive. This rarely happens. Most suppliers who have suffered from traditional purchasing regimes of the zero sum type are usually inquisitive to find out how they could benefit from this new change in policy. Ironically, some supplier businesses that may compete will be quiet and will need a 'code of conduct' if they are to engage in a meaningful way but this can be approached on a case-by-case basis.

8.3 ALIGN, VISUALISE AND IMPROVE

The first phase of supplier engagement is to visualise the future – the role of the policy deployment process. Initially the stabilisation stage for the supply chain will focus on the quality and delivery of products. Gaining control of the on time in full delivery of materials reduces noise and variation for the manufacturing customer. Limiting issues with suppliers will help to improve the OEE of the factory through better quality products and importantly the availability of materials so there is less time lost to the production process. The supply process is identical to that of the internal implementation of Lean TPM – stabilise and then optimise.

To convince any supplier to change, it is important to provide them with a vision of the future, the benefits of investing in the customer relationship and investing in their own improvement process. If suppliers see benefits then they will also realise that they can exploit the Lean TPM system for other customers – possibly your competitors! That last sentence would have caused shock to some people. You are developing your suppliers and the same suppliers may service your competitors. But in behavioural terms it is unlikely that the improvements gained by using Lean TPM will help to lower a supplier's costs and therefore raise their profits so long as the supplier does not volunteer cost savings instantly to your competitor. Very few companies will do this and many will also be working to a customer-owned product drawing where they cannot influence the real product cost anyway! To some extent there are circumstances where it is actually preferable to let competitors know what is happening – take for example the introduction of a pull system based on returnable boxes with barcode labels on them. It makes sense to find an agreement on box sizes and also a standard label format SOP that critical mass can be gained and the supplier can benefit from a standard approach. What would be bad is if one customer uses Kanban and others send in uncontrolled and erratic orders – so some cross education is good.

So a vision of how the future supply chain will work and what is needed from suppliers is critical. Again the new vision can be expressed as a future state of performance – gap if you like. So in 3 years the target performance level could be to improve quality by 20%, to reduce lead times by 30% and a measure of collaboration/innovation. The latter could be the percentage of suppliers with the ability (internally or with other providers) to design or redesign products or indeed the investment of suppliers in standards of environmental compliance. Such an approach was taken, at Toyota in the early 1990s and by 1994 all the suppliers had achieved the ISO14001 environmental management system and research proved all suppliers had generated improvement benefits that saved more money than it cost to implement the system.

By expressing the challenge, suppliers can see the critical measures of success – so quality and delivery performance are highlighted as important and for some manufacturing customers these measures are also expressed as target numbers for achievement. The classic example of this was the use of target measures for every year of Nissan, the UK's NX96 programme with suppliers. Nissan shocked its supply base (those outside of Japan) by stating quality in terms of the parts per million (ppm) product defects received when the European industry was measuring product quality in terms of percentages. So Nissan stated a target ppm value to be achieved by all suppliers for each year of the programme. Same too for delivery – this was expressed in terms of a time in a given day plus/minus an allowance. So initially Nissan stated a set time plus or minus 2 h and over time reduced this to just a matter of minutes. By stating these targets it focuses the mind and it really stimulates ideas from suppliers. These are the targets that determine whether the customer sees your business as good or not so good. Furthermore, more enlightened customers will measure these performance indicators and in a way that stimulates change too. A monthly report on all supplier performance (shown as a graph) and with an indicator on each personalised report of where your business is performing does stimulate effort. So let us say the average ppm level for all suppliers is recorded as 250 ppm and your business is at 500 ppm then you need to put more effort into the quality of product received at the customer. Initially this could be to contain the product losses so they are not shipped but eventually you will have to get to grips with the process – Lean TPM.

So the investment in better quality means the ppm defect level will drop from 500 to let us say 300 ppm in a matter of a month by sorting out dispatched product. That very same month (and the new report on all supplier performance will be available) the supplier average may have dropped to 200 ppm – because all other suppliers have not stood still! So this way the supply base in its totality is improving but importantly the poorer suppliers (who still have the capability and right attitude to remain a supplier) will make up ground. To master better and better the levels of performance, the poorer suppliers will expect some help and this comes from the manufacturing customer as well as other suppliers who have skills and techniques that could be shared – after all the suppliers do not tend to be competitors but rather businesses that will benefit from the entire supply chain achieving its collective goals.

But a picture says a thousand words and therefore it is important to visualise the future and current supply chain design. Again we have seen this tool earlier as the internal Lean TPM programme was undertaken. Extending this knowledge and training suppliers as a group makes maximum sense. It will also generate an action plan for every supplier that covers the different stages of change at the supplier business. As a chief accountant at the supplier company, this value stream map can be used to model future revenues and costs of supplying the customer and for the operations and maintenance staff at the supplier it signals the rate of improvement needed to supply the right quantity of product every day or week to the customer and therefore the options of decoupling the business and its supply chain.

Decoupling production is a very important supply chain design process and allows stock to be positioned in the supply chain to protect flow, to protect the availability and variety of materials that are ready for immediate shipment and also how replenishments are controlled. Initially most supplier businesses will decouple at the point of finished goods and this point is the point at which a customer can pull product into their system and the point at which the product is pushed to a stock point that protects the supplier operations. By engaging with Lean TPM, process robustness improves and it is possible to move the decoupling point inwardly to say a point where products are assembled to order (and not stocked in high quantities in finished goods). Some finished goods will be there but the mode of supply will be 'assemble to order'. Again with increased Lean TPM roll out, the decoupling point and main focus of the business could be moved to a work-in-process level so products are manufactured and assembled to order (materials are now spending less time delayed in stock locations and more time moving towards the customer – cash flow again!).

Lean TPM and the learning it generates, feeds back into better product design – so design for assembly and design for manufacture become realities and programmes with pivotal impact on the business. Take laptop production – most of the laptop is the same regardless of model – so what varies is what the customer wants to add in. Most 'add ins' such as sound cards, accelerators, USB points and software can be added during the 'assemble to order' stage – so design for assembly becomes a major competitive weapon. It can enhance flexibility, reduce the weight of the product and eliminate stock. Rarely does transformation programmes result in such knowledge (useful knowledge) get returned to product designers but Lean TPM is a massive exception to this.

Not just this but Lean TPM is a major support for the capital expenditure (CapEx) process for suppliers. The supplier network approach and value stream maps allow the customer needs to determine the designed cycle times and quality levels of new technology. Knowledge of the failures of the existing technology also provides new insights into what needs to be 'designed out' when the specification of new assets is undertaken. Such support from a customer sensei or vision of the future supply chain is invaluable. Inexpensive to deliver in terms of the discussion a customer will have with a supplier and all suppliers, but invaluable in terms of de-risking future investments. CapEx is again an activity that should generate future cash flows but it also has to be funded and supplier businesses will be interested in the 'pay back' period needed for any future asset. A supplier network that is actively meeting, discussing strategy and best practices will lower such risks and harmonise working practices during the stabilisation phases and then serve as innovation meetings as Lean TPM is extended.

There is also a mistaken belief that such supplier networks should be attended by the sales person and relationship holder. This is a falsehood. A team approach underpins the network and therefore operations, maintenance, design staff and team leaders of production cells should all, at the appropriate times, attend such meetings. The building of relationships and sharing of knowledge is all about the community of practice and how Lean TPM methods are being deployed and the learnings captured.

What can be achieved?

A recent case study presented at AME, Toronto set out an example of supply chain optimisation involving the coal supply chain for steel production. The route from mine to steel mill involved four companies optimising their own operations that are listed below:
- coal production (5 locations);
- rail transportation (600 miles);
- port management (3 ports); and
- shipping lines (multiple).

The 'system' that linked these organisations incurred significant waste: what could go wrong, did go wrong; push instead of pull, demand amplification, unforeseen (but predictable) problems at organisational interfaces.

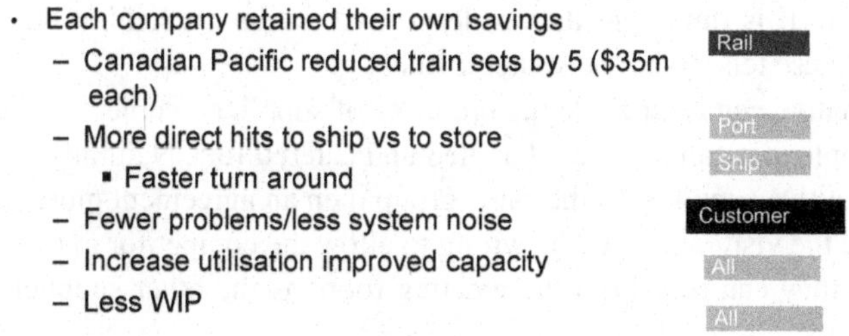

- Each company retained their own savings
  - Canadian Pacific reduced train sets by 5 ($35m each)
  - More direct hits to ship vs to store
    - Faster turn around
  - Fewer problems/less system noise
  - Increase utilisation improved capacity
  - Less WIP

Rail

Port
Ship

Customer

All

All

FIGURE 8.4 Gains from E2E supply chain collaboration.

By getting together and gaining an insight into the real causes of these wastes, they were able to understand the potential gains of collaborating and develop the trust to adopt E2E lean principles. These were the following:

- trains to schedule movement at regularly spaced intervals;
- standardise rail car type and length;
- plan to maintain a steady constant volume at the mines and adopt a first in–first out sequence at the ports;
- built in plans to recover operations after disruption;
- common performance metrics shared real time by all companies.

Precise benefits cannot be shared but the anecdotal gains were mouth-watering as set out in Figure 8.4.

8.4 SUPPLY CHAIN IMPROVEMENT SUSTAINABILITY

There is a decay curve for supply chain improvements teams as much as there is a decay curve associated with the pleasure gained from going from making lots of improvements in the short term to making fewer improvements and taking longer to reach that improvement.

The extension of Lean TPM via a supplier network means that a number of design decisions must be undertaken and ground rules established. These include the following:

- How frequently the group will meet. It is important that the team of suppliers meet regularly rather than sporadically. Issue-specific groups can be established but the main group should meet at regular intervals throughout a year.
- The people engaged with the network should be established. Managing directors of supplier businesses are busy people and therefore probably cannot spend their time attending every meeting. So it is important to have regular strategy meetings (at least three times a year) and then other meetings that look at operational issues and Lean TPM techniques where the managing director does not need to be present and alternative staff will find it of more benefit. So the meeting agenda needs to be set in advance and a routine established.
- The venue for the initial network meetings is typically the manufacturing customer of the group and this allows suppliers to meet key contacts at the customer, tour the facility and see how their supplied products are used in

practice. It is important that, at least, two events are held at the customer site – these tend to be the strategic ones!

- Alternative venues include the premises of suppliers (subject to the number of people that can be safely handled and catered for). Naturally, if there are competitive suppliers in the same group then an agreement must be reached before the visit – it would be wrong to allow the competitor to tour the facility so they can remain in the meeting room as the other suppliers tour for example.

As a learning group, the supplier network is a very effective and valuable activity. Internal Lean TPM programmes at customer businesses will be subject to OEE measurement variability if supplier product quality and availability of materials reduces optimum operating time for the customer conversion process.

8.5 SUPPLY CHAIN ENVIRONMENTAL SUSTAINABILITY

The credibility and brand reputation of a customer can be tarnished in a second. The world is far more transparent and this is particularly true for the subject of good corporate citizenship. Some fashion companies have been singled out because they have used child labour in Third World countries to make luxury products for the West. Environmental considerations rank equally high and the Lean TPM programme may seem a bit removed from environmentalism but this is not the case – Lean TPM saves carbon footprints and supports a 'green' supply chain. Fewer defects that cannot be reworked and must be land filled, less cutting fluid used, fewer losses of neat oils, less heat loss, less compressed air loss and many additions to the sustainability of the factory and its supply chain are critical and evidenced examples of good practice. Furthermore, Lean TPM is a catalyst and major input to better product design. Such a link is crucial and creates flow of knowledge that allows the production system in the future to be optimised. In the future, the product will design out inefficiency and costs of production and to the environment are minimised – all that before a single new product has come off the production line.

8.6 SPLITTING AND SHARING THE GAINS

Lean TPM provides a lot of answers to employees who ask 'what is in this for me?' in terms of new skills and a new value to the organisation as well as providing a working environment that is stimulating and challenging. The same question will be asked by suppliers – 'what is in this for my business?'. Yes, there will be protection of business between the companies because the customer organisation will be reluctant to replace a well-performing supplier even if a competitor was to offer a lower piece part price but there are other incentives that can be applied – these reinforce the value of 'collaborate to compete'.

The best example of this is the use of A3 improvement projects with suppliers. Gains made by the two companies by targeting key opportunities to improve will need to be shared in some way. So let us say the supplier finds a way of reducing the costs of a product by substituting a costly material by a cheaper alternative that provides the same safety and performance features. The options are as mentioned below:

1. The supplier keeps the saving and uses it to reinvest in more improvements. This is by far the simplest approach and the gains can be built into the next generation of product that the two companies will make.
2. A sharing agreement is put in place that allows the companies to split the saving. This can be complicated because some cost savings do not relate directly to a piece part saving or that when applied to the piece part price then the saving is tiny.

We have provided this advice here because it is relevant and will be an issue unless it is addressed at the beginning of supplier development activities. In reality the best approach is to simply let the supplier take the savings and to promote the approach to improvement and resultant techniques to the rest of the suppliers. Success begets success and more improvements will follow. If the system is seen to be 'one-sided' then supplier engagement will be limited and motivation to improve reduced.

It is far more powerful to reinforce the message of improvement by ensuring that improvements lead to customer operational benefits and that the supply base gets the clear message that the quality of improvements (targeted improvements) is more important than the quantity of improvement (Kohn, 1999). Also that the improvement should be derived from the value stream map (VSM) and therefore the improvement can be drawn onto the map to make it a living document. This living document becomes highly important when the supplier and customer organisation work together on a one-to-one basis. The VSM is the one document that focuses improvement.

8.7 TYPES OF IMPROVEMENT

The vast majority of improvement activities require little or no expenditure in order to generate benefits – they require an investment of time and a common method of getting to the root cause of a problem. The following list is a few examples of what typical networks focus on:

- exchange of economic, materials, competitor and business strategies;
- the introduction of management standards such as ISO standards covering quality management (9001), environmentalism (14001) and others (EFQM, Shingo Prize, TPM prizes);
- the levelling of demand and introduction of the Kanban pull system between companies;
- the introduction of a common logistics carrier that picks up from different suppliers on its regular route to the main manufacturing customer;
- lean techniques such as visual management, quick changeover, leader standard work, design of experiments;
- discussion of important issues that will affect future business – such as new labour laws being processed by legal bodies (for example, the Working Time

Directive being passed through the EU would have attracted the attention of such supplier groups;

- methods exchange such as barcoding, value analysis and value engineering and human factors;
- methods for better energy management and reduction of carbon foot printing; and
- customer presentations, network member presentations and guest speakers on subjects such as 3D printing, composite materials, new thinking on leadership styles, lean accounting or exemplar businesses involved with IT.

The art of keeping the network engaged is to remember that this is predominantly a relationship development activity that must result in benefits or it will become a purely social activity. It is quite common to find the meetings are conducted on the day after the suppliers arrive at a local hotel to enjoy an evening of networking, a speech and a meal. The next day is fast paced – no long lectures – lots of activities and break out teams to go and discuss key issues. It is also common to find that 'anonymous' questionnaires are used to assess the meetings, the pace of change, number of ongoing projects and such like.

The Lean TPM programme and the integration of suppliers is a major addition to the competitive arsenal of a business. It begins when sufficient internal progress has happened that the focal customer business to tell and show suppliers there is a better way to manage. This is far removed from a traditional programme of 'demanding' suppliers do something to improve and the network is the point at which the business recognises its dependency on its suppliers if it is to achieve its strategy and deliver on its promises to the consumer. This is a very common sense approach to business but sadly common sense is not quite as common as it should be.

8.8 CHAPTER SUMMARY

Supply chains compete. Most markets do not have a wide range of alternative suppliers and instead there are a few suppliers per product. Aircraft, cars, writing pens, coffee, furniture and many more sectors are dominated by just a few companies. The performance of these companies is partly determined by the business but also by the combined efforts of their suppliers. Just one poorly performing supplier will stop an entire supply chain. So it is vitally important that businesses select suppliers based on their quality, delivery, flexibility, capability to innovate and their positive attitude towards collaboration. It is now a question of dealing with the right type of supplier – a supplier that will learn and improve so that operational excellence is spread and suppliers share best practices amongst themselves and engage in the policy deployment process.

Improvement conducted on a supply chain scale improves its robustness and resilience to shocks. It increases dependency and collaboration in a manner that is focused on growth and, as the Toyota supply chain is probably the best example, that collaboration leads directly to diversification. So, as Toyota entered the markets of forklift trucks, sewing machines and house building (as just a few examples) the legion of suppliers diversified too (Liker and Meier, 2008). So Denso took its air conditioning knowledge of automotive products into new applications – such as the air conditioning of houses.

So supply chains should evolve – they are really designed and they represent a new challenge to the organisation that has learned the process of Lean TPM internally.

REFERENCES

Belcher, J. (1987). *Productivity plus: How today's best companies are gaining the competitive edge*. New York: Gulf Publishing.
Goldratt, E. (1984). *The goal*. Aldershot: Gower Publishing.
Kohn, A. (1999). *Punished by rewards*. New York: Houghton Mifflin Publishing.
Liker, J., & Meier, D. (2008). *The Toyota way fieldbook*. New York: McGraw Hill.

Chapter | Nine

Sustaining the Improvement Drive

9.1 INTRODUCTION

Implementing Lean TPM is certainly no easy task but it is achievable with hard work, application and making the implementation process something which employees and later suppliers find enjoyable and rewarding. It is after all a learning journey. Each stage of the Lean TPM journey includes an element of sustainability – like a ratchet system – whereby gains are secured before moving on to the next stage. The development of a sustainable business improvement model is important if gains are to be exploited and the workforce is to become empowered to engage in self-directed improvements. To this end there are a variety of techniques that allow the self-initiated improvements, sustainability of workplace improvements and factory-wide change processes to take place and sustain (Liker & Meier, 2008; Standard & Davis, 1999).

Sustaining improvement activity lies at the heart of being competitive and this most difficult of activities has two dimensions. The top-down dimension concerns the processes and techniques that can be used at the management level and deployed via the policy deployment and Lean TPM master plans. The bottom-up dimension is one where knowledge is transferred and empowerment increased. Both require a spirit of collaboration.

All 'world class' manufacturing businesses have a team of committed managers who share a common vision of the future and work together, via the policy deployment process, to achieve it. These organisations know that collaboration between managers is preferable to conflict or prioritising the department over the interests of the company (Belcher, 1987; Sobek & Smalley, 2008). Managers control large parts of the manufacturing and supply chain, and these individuals occupy positions in the business whereby their decisions can have profound transformation on the flow of materials throughout the value stream.

To put it another way, several good decisions taken at the management level can generate large changes that release millions of pounds of benefits to the business. But these benefits are only held if the Lean TPM process is present. Stable and optimised processes hold the gains and translate improvements into commercial realities.

9.2 SUSTAINABILITY AT THE MANAGEMENT LEVEL

The sustainability of the Lean TPM approach, at the management level, involves a series of mechanisms which demand integration, learning and consensus building. These are important processes but they take time to develop and must be deliberately scheduled into management meeting time especially where the company managers share a history of 'in-fighting' or 'power gaming' (Peters, 1992). These processes are supported by three types of improvement tools: (1) frameworks to make improvement potential visible, (2) change processes to deliver new ways of working and (3) modelling to reinforce new behaviours and working relationships.

Traditional improvement tool training tends to concentrate on the first of these covering framework tools such as mapping or problem analysis. The other two tools change processes and modelling are not covered. Expecting this alone to deliver improvement is the equivalent of expecting to lose weight by simply standing on a set of weigh scales. Mapping on its own is never enough.

The preconditions for sustained improvement include competence in both change processes and modelling behaviours. This ensures that the Policy Deployment process becomes a means of engaging managers, as a group, to vision the future and establish a gap with the 'current state' product value streams. These activities allow a focus on 'customer value' to be combined with an understanding of where the company is now thereby identifying the 'value gap'. This 'gap' is therefore the challenge that must be closed if the business is to remain viable. No single business department has all the answers with which to bypass these two processes and therefore in designing a current and future state, virtually all departments will have to contribute to a greater or lesser extent.

From the value stream mapping stage it is possible to identify a wide range of improvement initiatives in each department to improve performance and move the business closer to its 'future state'. These improvements tend to be tactical issues that need to be addressed as they are known or identified weaknesses at the time of the mapping (they remove the barriers that affect today's business). The next stage is, therefore, to take a view of 'what the customer and business needs' over the next 3–5 years. This time period, for determining customer value and also what business stakeholders expect, has many advantages to the modern business.

The policy deployment process is guided by the Lean TPM master plan. This sets out the major transformational steps on the journey from reactive to stable to optimised operations. This operates at the 3–5 year planning horizon. This is sufficiently long enough in the future for any capability 'gap' to be closed by the collective action of management but it is not too far ahead to be impossible to predict what is wanted from the business.

Focusing this management discussion throws up many interesting interpretations of the future and tends to gravitate to discussions concerning new products but more importantly measures of operational performance expected by customers. These commercial discussions, including benchmarking of competitor performance, tend to result in a series of challenges that face the business. Generally these are expressed as a quality improvement target, delivery reliability and lead-time target, and finally a cost-reduction target. As most lean manufacturers know, these targets have a common logic and by improving quality delivery reliability, lead times and costs tend to fall. So the emphasis is most definitely a reconfirmation of 'quality' improvement throughout the business. As these targets and commercial measures pass through every business department (including maintenance engineering) to yield business performance, most lean businesses will deploy these challenges to the entire group of middle managers. Middle managers, or departmental managers as they are often called, know what is best practice in their specialisms and therefore they are best placed to find the 'hows' or key projects needed to meet these goals. They also know the changes that must take place within the organisation to deliver these improvements but, under traditional thinking, could not influence these changes.

However, when these middle managers are put in an environment where they are challenged to find ways of meeting the key business challenges, it creates a forum for discussion, learning and also the identification of key improvement programmes. The forum also provides an ability to put into perspective the best practices of each department within the context of the overall business performance improvement required. As such the middle management forum allows certain projects, of low commercial worth but regarded as functional best practice, to be delisted or accepted as a good initiative. Add to these deliberations, a finite resource of money or employees and the key initiatives at the departmental and cross-departmental levels will emerge. Having identified these programmes of change it is important to combine them, identify the resources needed and finally to identify a timeline for implementation. There are many techniques that add to this aspect of sustainability and one of the easiest is the policy deployment 'X' chart form (the name given to taking a set of challenges all the way through to action plans and review processes).

The policy deployment X chart

In Chapter 3 we introduced the X chart for policy deployment, and in Chapter 7 we introduced the X-type chart for product quality attributes and their implications for optimised production management using Lean TPM practices. The policy deployment X chart provides a visual chart of business and operations management improvement programme selection. The chart demonstrates the areas in need of improvement and the tasks needed to achieve the business step change in one simple management chart. The format also allows the actual tasks of the improvement activities to be assigned to teams (with a leader and team members) and define the timeframe over which the team-based improvement activity will take place.

To recap, the X chart is a logical and graphical display that shows the following elements of a focused change process (Figure 9.1). The chart commences with a listing of the key elements of the cost equation in the form of areas of waste. (Box 1) and then the specific target waste area for improvements that the management team have identified as key projects to influence and reduce the total costs of business failure (Box 2). From here the management team determines the "improve to" targets (Box 3) and then the expected benefits that achieving these targets would realise (Box 4). Finally the sum of all the items in Box 4 is

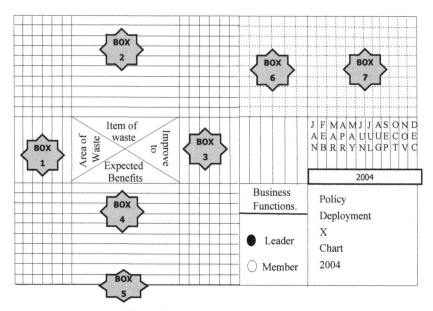

FIGURE 9.1 The policy deployment X chart.

presented in Box 5 to declare the overall business savings possible as a result of undertaking the identified initiatives. Returning to Box 2 and the key projects – the far right of the chart shows, for each project, which cross-functional business departments should be involved. It also shows which department has been assigned as the leader of the project and who else will support them. The cross-functional teams are shown in Box 6, and, to the right, a general project timeline showing the start and finish of each project (Box 7). The project timeline allows, on a single piece of paper, every project to be seen, its contribution and when it is expected to deliver its value stream improvements.

The illustration chart just simply shows a couple of projects concerning the management and reduction of 'work in process' costs (Figure 9.2). Here two areas of waste have been identified for improvement, batch sizes and setups, and the target is to halve the batch size and also halve the set-up time. Looking across the chart, the set-up time reduction programme will be led by operations and supported by the technical department. These two groups of employees will begin the programme of improvement in February and finish it in July 2004, and if they achieve their goal then this will contribute £250,000 of savings for the business. If this round of improvement ideas is continued and all areas of factory waste are identified, then it is possible to identify huge savings resulting from only a few projects. As all managers are involved each will understand the purpose of the project and how they can use the results of the activity for themselves. This is a truly valuable exercise for management; it delists those projects that will add no value (or are 'wish list' items for specific managers) and will confirm the major programmes of change needed at the factory level to add more value for the business and its customers.

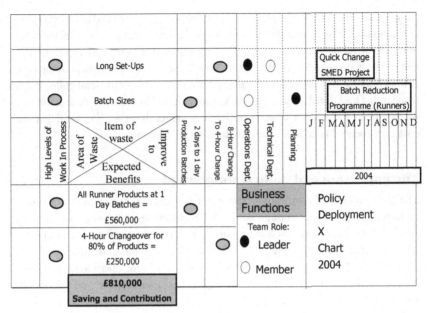

FIGURE 9.2 Illustration of a business-level X chart (showing projects in operations).

Having established the key change programmes that have been determined by the cross-functional team to optimise the value stream, the necessary project management routines need to be introduced. These plans are critical to the business, and it is important to understand any linkages between projects that will yield the new production system design (the critical path). At this point, the cross-functional team now has all the analysis and countermeasure projects needed to 'focus' the implementation process and to root out the areas of waste that prevent optimisation of the 'runner' value stream – the one with the most profit potential if redesigned effectively.

The chart above shows the duration of the project over the coming year or so and identifies who will act as the senior management champion, the operational change leader and the key departments that will form the implementation committee. The latter cross-functional group ensures that all the major stakeholders engage in the change programme and that the 'Lean TPM' initiative is not seen as an 'operations project'. This is an important feature of the optimisation process, and this structure offers a very powerful platform upon which to optimise and to identify and lower the risks of any given course of action. Furthermore, it also integrates the operations department with customer-facing and supplier-facing business departments in a manner that is focused on exploiting the 'value' of the manufacturing process (for market effectiveness) as much as generating high levels of efficiency.

Iterations of the annual improvement planning process will incrementally identify new levels of organisational waste that constrain the value stream and they will also identify trends and shifts in customer expectations. These are important to ensure a sustainability of process improvements towards the 'zero losses' or 'waste free' optimised state of production. The 'Lean TPM' approach has a natural complementarity with the needs of other business departments and also prevents the promotion of TPM as merely the concern of the maintenance of production functions and increases the learning across all managers. The latter is an aspect of TPM that is poorly managed in the West even though a successful TPM programme needs a 'total' buy-in from all business managers if it is to yield commercial gains. It is no wonder that traditionally these process improvements have failed to deliver the commercial benefits required when they typically started from an internal rather than customer-focused approach to design the ideal production system.

The importance of daily management

The core process of reinforcing new ways of working (modelling) is the daily management process. Although we have developed separate processes for decisions about short-, medium- and long-term horizons, the factors which impact on success or failure of these processes occur every day, at the same time. They need to be managed simultaneously. Ideally we want the routine operations to be stable so that the customer order fulfilment engine becomes a self-regulating system. Unfortunately, although we do not know the precise changes in customer needs, we can guarantee that their desire for quality, cost and delivery will be travelling in one direction. That means we need to be constantly refining our approach to day-to-day operations. In addition, the growth we seek is most likely to come from new products and services. Not every new offering will succeed and in many industries, there is first the need to fill the pipeline. If the product is successful, the pressure on operations can be intense. If the product is not successful, with full warehouses, the operations can suffer from lack of space. The daily management process needs to be able to take all of these steps in its stride.

In this environment we need to make full use of visual management. Not as window dressing to impress the visitors but to
* make important information visible at a glance,
* clarify the current reality and provide the insight to better manage the future.
The effective use of visual management also
* places knowledge in the public domain,
* reinforces ownership of the territory by its occupants,
* encourages work by small groups and improvement teams.
(Source: Visual Factory, Michel Greif: Productivity Press A study of visual management practices at 20 + French/USA Plants).

In this environment we also need managers to be consistent in the way they manage the routines. That way, each manager, each shift will take day-to-day decisions in a way which reinforces the context of an agreed long-term plan. Figure 9.3 sets out a summary of the areas of competence needed from first-line managers to support this. This covers not only understanding of the day-to-day shop-floor reality but also the links between policy deployment (tactical) and business strategic goals. It also includes competencies to be able to develop high-performance team capabilities. This is supported by the development of management standard work to simplify and delegate routine management to front-line teams as they develop capabilities.

- **Know the territory you manage**
  - Understand the numbers
  - Be a coach, help the front line to do their best
  - Give recognition
- **Touch the process**
  - Strategic vs tactical
  - Block out time to visit the shop floor, set the cadence for conversations
  - Incorporate shop floor visits into management standard work
- **Establish visual management standards**
  - Visual controls to identify "normal" conditions at a glance
  - Metric boards to know the game plan, understand the status, confirm the tempo from 50 feet and in 5 seconds
- **Use the daily management system**
  - to trap problems early
  - To reinforce standards
  - to follow up to prevent reoccurrence
- **Develop management standard work**
  - Add the value that only you can add
  - Increase the effectiveness of value adding time

FIGURE 9.3 What we want from managers.

5. Transformation Process	100% engaged in teams to deliver new value and step-out capability	Driven by Vision
4. Improvement process	Over 50% of people engaged in teams targeting waste and focussed improvement as part of a top-down and bottom-up improvement programme	
3. Cross functional improvement team	5–10% of people engaged in top-down-driven cross organisational initiative	
2. Functional Improvement Team	5% of people involved in improvements based on functional priorities	
1. Ad hoc	Individual allocated a problem when one occurs which is painful enough.	Driven by Pain

Sustaining Improvement Behaviours

FIGURE 9.4 Front-line team capability development steps.

This development of team capabilities systematically transforms the drivers for improvement from one of dealing with problems to one of releasing potential. It is this increase in front-line capabilities that characterise the transformational steps of the Lean TPM master plan as set out in Figure 9.4.

The daily management process therefore combines completion of day-to-day operations in a way that engages managers and shop-floor teams in conversations and activities leading to increased empowerment and operational capability as set out in Figure 9.5.

The underpinning systems that integrate this daily management process into the policy deployment process are set out in Figure 9.6. This shows how the

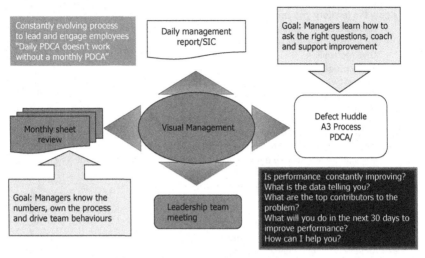

FIGURE 9.5 The scope of daily management.

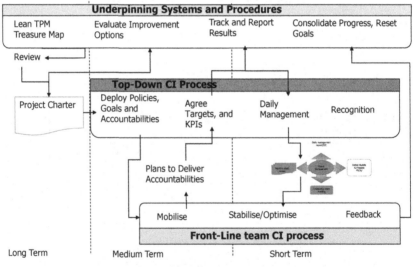

FIGURE 9.6 CI underpinning systems scope.

top-down policy deployment and bottom-up daily management agendas combine to provide tactical coherence and strategic control. This diagram sets out the elements of the company-wide change management process which needs to be mastered to secure never ending improvement.

Contemporary management and leadership

The impact of daily operations management behaviours on the sustainability of improvement programmes cannot be overstated. This behaviour determines and reinforces the culture of the firm. Any manager who therefore pours scorn on the

TABLE 9.1 Old and New Management Approaches

Traditional Business Behaviours	Lean TPM Behaviours
The business structure resembles highly demarcated functions with rigid hierarchy of responsibility	Business is oriented to value stream (product family management). A flatter and less layered organisational structure
Decision-making is centralised to business functions	Decision-making is decentralised to the value stream team
Managers plan, instruct and control	The manager is a coach to the team and develops the team to take on responsibility for processes and their improvement
Support departments hold all the specialist knowledge in the factory. Knowledge is guarded and not shared	Specialist knowledge is passed to the team and this transition is overtly managed to allow specialists to join the team or move to higher levels of diagnostic project work (higher value added)
Rigidly written and highly defined job definitions and grades	Broad job descriptions that increase flexibility of employees. High levels of empowerment
Employees engage in task activity	Employees should understand the process, in which they work and should be multi-skilled
Problems are referred to specialists	There is no divorce between identifying and solving problems. Employees are appropriately skilled to solve problems to a high quality of solution
Employee focus is on processing work as fast as possible	Employees process work as fast as possible given safety and quality requirements of the work and in line with demand for the work
Focus on cost reduction	Focus on cost reduction through quality improvement and delivery improvement
Inspect in quality	Design in quality and inspection by all

process of improvement is likely to be heralded (especially by those who seek to slow or hijack the Lean TPM programme) as an example of why employees should ignore the programme and revert to type. Devaluing the skills and competencies needed to fully exploit a Lean TPM programme is a sure way of halting its progress as too is the wanton avoidance of leading change in the factory by those managers. The lean approach therefore needs a personal mandate from the entire management team to move from a traditional organisation to a lean business (Table 9.1). Progressing from behaviours in the left-hand column to those on the right of Table 9.1 is the behavioral modelling challenge which must be met to progress towards industry leading performance. This is the final strand of competency development incorporated in the Lean TPM master plan steps.

Therefore a Lean TPM culture must be developed to complement the main improvement and implementation programme. All too often manufacturing businesses ignore this cultural dimension and therefore fail to identify problems with people and miss out on the additional benefits of a culture that accepts change. These issues which will inevitably be faced by managers include a wide range of behaviours which need to be displayed in the factory and written into factory policy to guide the behaviour of managers at every occasion. Also these policies set a framework for dealing with people – they supplement the formal contract of employment – and can easily be implemented with a little thought. These issues and aspects of a sustainable Lean TPM programme are shown in the next table. Each of these factors requires thought to ensure that intent and reality are matched but no issue is a formidable task and most can be solved with a (or series of) simple solution.

One of the most important aspects of sustainable change within the firm is the regularisation and quality of information passed to the workforce in offices and also at the operations. Even the business costs of replacement parts, lost working time and other financial information should be deployed to these teams. No team can ever know enough about the firm if it is to make the necessary and appropriate contribution to the business. As such these simple issues – the ones that 'close out' management thinking are important to operation-level employees. Each of the countermeasures introduced should therefore reinforce the basic values of the firm – that of growth, development and improved value adding. In this respect, many world class businesses feel comfortable with writing down and displaying throughout the factory (and on the business cards of all managers), the formal management policies of the firm. This is good promotion and also serves to 'police' the factory management team. By doing so, it becomes obvious to the workforce when management is backtracking and must explain why they have been perceived to breech these fundamentals of improvement.

The public changes in behaviour and changes that have resulted from Lean TPM most significantly the reduced firefighting and cross-functional management processes have created a dialogue about the business and a focus on the business – in effect a systems rather than departmental approach to management. Reflecting upon the management journey, the modern manager has shifted significantly from a traditional definition of 'manager' and even the definition of what a manager did under the pre-stabilisation stage of Lean TPM has changed too.

The new behaviour is strategically aligned and less time is spent being a 'control jockey' chasing people around the factory or remotely monitoring staff with complex spreadsheets and pivot tables. The focus of being a manager is on the business and what is best rather than the stress of fighting a departmental battle that no one can actually win! Under the pre-stabilisation stage, a manager who was not firefighting and spent time thinking about how to improve was often considered not to be doing their job or running away from their responsibilities to dictate change and yell at staff. The new continuous improvement process crosses boundaries and is fun rather than early lean where it was focused and determined by managers. Effectively the early lean stage uses a team approach to randomly kill off sources of failure – but this random and departmental approach, often meant an improvement in one section of a factory, had little or no measurable impact on the business – it just solved a local problem. Typically this would speed up product flow at that point in the process only for the product to sit in stock at the end of the process for weeks! If you are at this stage of your lean implementation it is best not to say this out loud. The modern

TABLE 9.2 Principles and Demonstrated Application

Key Principles	Displayed by	Why?
Trust and integrity in the factory	Management behaviour and joint working groups	Basic condition for workforce collaboration and respect (not necessarily admiration or liking of management)
Mutual respect for co-workers	Secondments to project work. Joint training sessions. Involvement of indirect functions with operations (including sales and marketing)	Basic condition for collaboration
Openness	Increased use of visual communication within factory. Meetings with appropriate data resources available (not opinion)	Basic condition for collaboration
Entrepreneurial behaviour of employees	Management–employee briefing sessions and feedback	Enthuse workforce to feel they can contribute to destiny of factory. Align thoughts to business level not function
Flexibility of individuals	Broader job descriptions and more-salary-style payments	Less demarcation as a barrier to involvement or rotation within factory
Customer and market orientation	Communication to workforce concerning business and market conditions	Promotion of 'value thinking' and benchmarking current activities with a commercial output for the firm
Teamwork and collaboration	Problem-solving and mapping of processes by cross-functional groups. Common brainstorming sessions. Learning and reflection group time	Collaboration rather than competition or adversarial relationships
Less demarcation between employees	Single company uniform, common canteen, common car parking. Wider job descriptions	Improved flexibility of workers and feeling of legitimate role and involvement with business
Technical competence	Logical training of employees in skills needed to conduct tasks and problem-solve	Basic level of management efficiency and effectiveness. Means of deploying decisions down the organisational hierarchy
Shop-floor (bottom-up) initiative and involvement. Ownership and autonomy	Performance boards in factory areas. Suggestion process to 'flag up' issues Clearly marked boundaries and role of each team Self-managed problem-solving	Decentralises decision-making and stabilises production system by eliminating/identifying errors

Continued...

TABLE 9.2 Principles and Demonstrated Application—continued

Key Principles	Displayed by	Why?
Identification of abnormal conditions in the factory. Quick resolution of problems. Sensitivity to safe working practices	Training in area and asset conditions (normal operating conditions). Auditing of factory standards Visual management Regular problem-solving to capture any slippage	Quicker resolution of problems Feedback concerning skill shortages in identifying and solving problems
Formality of standard operating procedures	Written standard operating procedures Single point lessons	Basis for improvement so that 'reinventing the wheel' is stopped. Teams build from each other and can share good practice

approach is future-focused and it crosses boundaries – so why not supply one of your production team leaders to go map out the recruitment process or the end of month accounts reporting. Crossing lines on the organisational chart is OK because they do not actually exist in reality and they get in the way of flow if you focus on them rather than the value adding process and its flow.

In early lean – during the stabilisation phase – the focus is on doing things right and now the business looks to do things differently. This is characterised by a focus on consistency of principles and values such as those set out in Table 9.2. Hard measures are also used but to improve learning. The future is unpredictable and measures are of limited value where there is uncertainty. Values provide a better guide to the direction of true north. The future focus removes the politics of today and the pain of making changes because someone else has told you to (early lean). This is a major step forward and a massive improvement in the business and working life of all employees. More than this though – managers now listen. Managers do not have to have all the answers and be stared at when a problem arises for a mandate and a finger to be pointed at the person who would have to go and change things. The team is now the focus and there is an expectation that the team will not bring problems – they will instead bring their research and analyses (using the structured A3 approach) together with the list of alternative courses of action and a preferred option that the team agrees upon. This is a remarkable difference in management – it marks the knowledge management phase of lean and the ability to think about business growth – even diversification and certainly through the growth of everyone in the business. Long gone are the days when a brain was not essential for working in the factory or with suppliers/dealers to enhance customer value and profitability. These new measures are also important – they mark another change in thinking.

Improving Performance by Measuring it and Creating a Common Language of 'Value'

It is often said that if you show a person how they are measured then this will dictate their behaviour (Goldratt, 1984; Kohn, 1999). Thus if middle

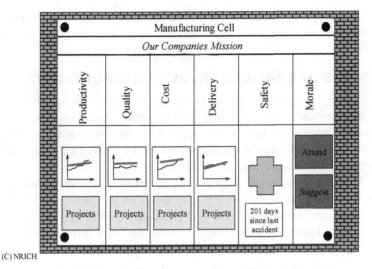

(C) NRICH

FIGURE 9.7 Visual performance board illustration.

managers share a common measurement system that includes the quality, delivery and cost of their departments then this is shared by every working area in the factory and appropriate measures of each can be found. It is no surprise that managers use measures to guide sustainable improvements within each department using these measures (and also 'safety' measures and 'morale' measures). In this manner, each area of the factory tends to have a large team performance board (in physical size) that shows line graphs of performance in each of these measures (more often with an annual target of performance for each, Figure 9.7). In this way, local area teams can use the board to focus their improvement activities and see for themselves the results of their efforts (as the line graphs change). Also these measures are unlikely to change over the years (instead the target measures will change but quality, delivery and cost will not). This form of sustainability brings together the manager and their teams to talk in a common language and to engage in improvement activity. These boards also form a bond between engaging in Lean TPM activities and witnessing improvements in area performance. At this management-led level, sustainability is personalised to the area of the factory and will provoke 'one to one' interdepartmental working where changes are necessary across shifts or by teams that supply products to the area.

The policy deployment approach and the common measurement system improve communication and add a level of focus to improvement activities that are related to the improvement of the firm. These improvements may also be allocated to individual teams and team leaders in the same manner as that of middle management policy deployment. However, these two elements provide a focus and direction to sustainable change. They are enablers if you like. The implications though include the allocation of time and resources to achieve these improvements – again a feature absent from traditional improvement planning and execution.

9.3 THE OPERATIONS LEVEL OF IMPROVEMENT

Most people work in the operations level of the firm and it is at this level that an infrastructure and careful attention to sustainable improvements are needed. Our discussion so far has really concerned the supporting practices that allow employees to add value and contribute to the firm. There are many important enablers to sustainable improvement at this level.

At the early stages of a Lean TPM programme, it is important to establish a communication and support mechanism for the factory workers. This is often termed a 'promotion office' and is basically a set of employees who plan, facilitate and guide the improvement programme. These individuals also control the standardisation of factory systems including documentation (especially the quality management system) and training materials. These individuals also have a key role in assessing the quality of the application of improvement activities.

With the role of assessor, the promotion office team has an ability to assist with the necessary training to upskill the workforce through additional qualifications. The most important qualifications are job-related qualifications that assist operators in gaining a deeper understanding of the processes and technology they control. The ability to combine work with qualifications, especially qualifications that are designed directly to meet the needs of the firm, provides an important source of sustainability. Furthermore, it is quite inexpensive as these life skills create a 'pay back' throughout the entire life of the individual's service to the firm. External training also breaks the monotony of the traditional factory environment and signals a clear message to the operations teams that the management is 'serious about change'. These courses also stretch over many months making it difficult to sustain the criticism that 'Lean TPM' is another 'silver bullet' or 'flavour, of the month. Again there are echoes here of the 'pleasure' of change reviewed earlier and rewarded effort.

Another responsibility of the promotion office is to act as a regulator of change (to pace the amount and timing of activities). The ability to control the pace of change is important to 'employee learning' and the achievement of certain standards in the workplace before new techniques will be passed to the teams. At Toyota there is a saying which states 'the teacher will appear when the student is ready', and this is not typical of traditional improvement programmes that teach everything to everyone and expect them to come up with solutions. The logic of the implementation process under Lean TPM is to pace change and reveal the next set of improvements and techniques when there is clear evidence that the last have been mastered. There is another important point to be made here – the promotion office and early stages of Lean TPM are concentrated upon the quality of team solutions and not the quantity. This is important as successful implementation is a learning experience and therefore exemplar innovations must be created (some of these will have a low commercial value but will stand in the factory as lessons for other teams). Only when the basics are mastered will effective promotion office direct efforts towards the quantity of solutions. Such a 'checking' mechanism stops teams from half-understanding the tools and techniques and setting off in a direction of their own in an uncontrolled manner.

Add to this approach to capability development the creation of rewards for effort, improvement activities and engagement in key policy deployment programmes and you can stimulate directly a 'learning culture' that is capable of sustaining performance improvements. All too often managers run away from the issue of culture change. It is seen as somehow the last card of resistance or the 'thing' that is most difficult to improve. This is a falsehood, culture is the result of behaviour and in particular it is the result of management behaviour. Cynical managers who sneer at improvement programmes will breed a culture of cynicism amongst their reports. Managers, who enthusiastically promote change and improving upon the achievements of today, will not just be happier themselves, but will create a culture that embraces change more readily. In this manner, culture is an output not an input and changing culture means changing the management approach. Think further, treat you trade union as if they are obstacles to progress and guess what they will be and as a manager you will highlight even the smallest degree of questioning as resistance and guess what eventually it will turn into resistance. Sustainability of improvements on the shop floor is therefore a function of sustainability of the approach taken by management. This brings a further point; sustainability of operational improvements is also related to the job tenure of a progressive manager. If there is a high level of management attrition then this will also affect negatively the treatment of 'improvement' by shop-floor teams. Basically if there is little faith that managers will remain in place to see out the improvement programme then factory teams will take less seriously the mandate for change.

At the operations level of the firm there is a portfolio of practices that support process improvements, and these range from giving individuals specific projects to assist their career development to simply acknowledging effort and saying thank you. Further positive reinforcements can be found in terms of offering factory 'certificates' for involvement or achievements following successful change initiatives. These are powerful mechanisms and 'reinforcers' of change, but so too, is the proactive management of getting individual factory employees, or even teams, to attend external training courses and to network with other local businesses. We have even seen complete employees, already good at process improvement, totally reinvigorated after a simple (and free) public visit to a Toyota factory.

Breaking down barriers to change (therefore unblocking improvements) can also be achieved by concentrating on human resource issues such as individual career development plans, succession plans and suggestion schemes. But sustainable improvement is also related to job rotation and getting employees (especially managers) to experience different parts of the production system. This activity extends the boundaries of people's thinking by allowing them to relate their actions to the improvement of other people's performance in the factory. The same is true of operator teams. It is not sufficient to 'bring people in working days' to conduct problem-solving between shifts. These individuals contribute the most when they act horizontally across the business (between teams), and therefore secondment or the deliberate rotation of staff assists in gaining this knowledge and optimising the value stream in a logical manner.

The logic of operation-level sustainability

It is astounding that the planning of a major improvement activity, such as Lean TPM, is as we have stated before the result of 'fag packet' designs and no real attention is put upon the timing and logic of such a change initiative. There is a definite and totally logical approach to the actual programme of events at each factory and this logic is pretty universal. Also within each stage there is a logic that is common to each phase of change.

The logic of Lean TPM implementation is as follows:

1. Create awareness in the factory of a need to change using a collaborative approach to analysing the business as a single system of material flow. This level of lean implementation is best achieved by creating a value stream map by middle managers and key stakeholders in the factory (such as the trade union). The purpose of this stage is not just to create a sense of urgency to gain agreement on what needs to be improved and how best to do it. The 'future state' value stream that results from this activity will inform the business as to how best to get these improvements, what programmes of change are needed and what support is required (including the softer side of employment we reviewed earlier).

2. Implement initiatives to raise worker morale and safety management practices. Here the CANDO programme of change is invaluable and whilst many texts advocate this as a later stage, we disagree. It is one of the first. It is the first visual sign that management is concerned enough about this latest change programme to engage workers in it (rather than just demanding their compliance with a system they did not build). For some workers, especially ones that have received little formal training by the company or have been a long time away from their school days, this activity is enjoyable and an immense learning opportunity. Furthermore, to implement CANDO successfully it requires an informal level of brainstorming, collaboration and problem-solving. So by doing these individuals gain experience and will be less frightened of class room-based problem-solving activities.

3. Implement formal quality problem-solving activities with teams. This second stage is superb and compliments the CANDO and safety management stage by adding the formal tools necessary to identify, analyse and solve problems. For most employees this stage will be seen as a direct extension of CANDO that is logical to them. For the business, any improvement in quality performance will yield better productivity and therefore lower costs. By engaging in problem-solving the 'collaboration and involvement' aspects of Lean TPM sustainability are refined and exhibited. There is one further feature here that many managers forget – by conducting 'cause and effect' studies the team will identify problems with the environment! These are failings with the CANDO system and can therefore be captured and corrected at this point.

4. Implement new delivery mechanisms. It will take even the most enthusiastic team a while to get to this point, and it is here that the value stream mapping outputs can be implemented at the factory. By engaging in CANDO and problem-solving (plus associated training and other techniques) the management has bought themselves time to get the data and perfect (on paper) the changes needed. Furthermore, the teams have techniques and systems that can be picked up, moved and transplanted in the new production system. At this stage the teams will be involved in establishing the correct procedures and responsibilities to manage pull systems or operate in a flow environment.

5. Reduce costs and improve flexibility of the factory operations. The final stage at the operations level is therefore to seek out the 'designed in' safety stocks and eliminate them. When companies engage in lean practices they tend, quite rightly, to leave slack in the system that covers for risk. During this stage, after the production system has stabilised, the sources of this slack are questioned. These are costs. Therefore using problem-solving techniques (especially affinity style diagrams) a causal relationship can be established between improving machine changeovers so as to release safety stocks. This is an important learning activity for the team and extends the tools available to the team and also the level of cross-functional involvement they will seek. Whilst improvements like set-up time can be implemented before this stage, it is rarely the case that they are regarded as willing additions to the team. More often you hear complaints that 'managers are just trying to make us work harder' and managers will complain that 'operators do not sustain the level of improvements they have made during initial changeover time reductions'. Both these statements exhibit a lack of communication and understanding. This level of change is therefore more potent when it combines a willingness to sustain following self-selection of the problem (not dictation by management).

These general levels are enough to guide the implementation process but within each stage is an important set of criteria that maximise the enjoyment and value of each. Every stage must start with a promotion activity to generate awareness and announce that the improvement is coming. This is only common courtesy and also allows some time to get the necessary people released from their 'day job'. Furthermore, even in factor with traditional resistance to change, it starts the 'rumor mill' going. People begin to talk about it, question it, ask if they can take part and all these behaviours help the change programme.

The logic of implementing the change is also quite straightforward. Teams must find a quality of solution to the problems they face (not quantity). There are many companies that believe in 'blitzing' problems using these events but it rarely sustains. It is the quality of the solution (even to a minor problem) that is more important – quantity comes from repeating the work and sustaining it. So quality of improvement first (focus and following the process to learn it) – quantity of improvements second (increasing the number of improvement projects). This approach also firmly states that the company is not interested in meaningless 'short-term' demands for this improvement activity to pay for itself – the repayments will come and they will come for many years if the improvement process is planned carefully.

At the event, the first stage is to generate awareness for the team by showing them good practice and the reality of the factory. Here it is vitally important, especially where certain employees cannot read and write effectively, to train in pictures (show the team!). This avoids problems with getting people talking. It is not that difficult either. Alternatively take the team to another factory or to sections of your factory that show good practice. At every stage of the improvement programme it is important to build this awareness of the need to change. These sessions will also flesh out what the teams believe the management must deliver to make the programme successful (more often than not its time and resources). The second aspect of sustainability is that of education in more formal terms which includes giving the teams the knowledge and tools to conduct the change and the approach that should be taken. The third stage is to implement the change and monitor its performance (before and after) and finally the team must meet again to formally review progress. The last stage is important and should not be treated as merely a 15-min 'washup'. Instead this process should take a couple of hours and includes a reflection upon what has been learned, changes that should be made for future events and also to devise the standard operating procedures needed to ensure sustainability when the team project is disbanded. Whilst the project will continue after the teams have conducted the stages, these latter activities assist in generating an efficient (minimal time and effort) set of processes that regulate the system after the improvement has been implemented.

These stages apply to all the activity phases from 'safety and morale' to the 'cost reduction and flexibility' stage and to every team. Obviously repeating the training in the factory with new teams will involve updating and improving the approach and materials used too. The final activity is for the team leader and promotion office personnel to develop a short report (no more than two pages) for submission to the factory management and concerning the improvement activity. This report should also include how the team leader expects to continue the programme and whether there were savings (including an agreed diary schedule of future events). These reports form a library of activity logs that can be accessed by other teams facing similar problems. Furthermore, managers who have thought sufficiently about this process will also keep a log of 'who has been trained in what' and the standard they have reached. These are typically displayed in the team area and the promotion office maintains a record. In this manner, trained and experienced individuals can inform others and help them out between formal training sessions (even individuals from other areas of the factory).

This logic of sustainability is therefore to adopt a 'layered approach' to learning so that the new learning points reinforce the last and are seen as a natural extension to what has been done before. Take for example, the follow-up of a workplace organisation programme (CANDO) with 'Cause and Effect' problem-solving. The Cause and Effect chart has an area of brainstorming related to 'the working environment', and this will pick up any problems that have not been closed out during the CANDO programme. Quick machine changeover programmes also depend upon good workplace organisation and will be seen, by operators, as an extension of the workplace organisation but this time to do with sorting out what is needed at the machine to change it quickly. Layering learning in this manner therefore reinforces what has gone before and is not seen, by the bulk of employees, as just another initiative plucked from the air or from the last conference the managers attended. Such an approach builds and accumulats knowledge in a logical order. You will also notice that these phases actively reinforce the common quality, delivery and cost language and measures of the firm. As such, team performance can be viewed by a cursory analysis of area performance boards and the local measurement systems in place. Obviously some companies will have effective systems in place already that cover some of these areas but there is no harm in 'refresher' training sessions to reinforce the importance of conducting these tasks properly. Most people work in the operations level of the firm and it is at this level that an infrastructure and careful attention to sustainable improvements is needed. Our discussion so far has really concerned the supporting practices that allow employees to add value and contribute to the firm. There are many important enablers to sustainable improvement at this level.

9.4 SUPPLY CHAIN SUSTAINABILITY

There are many other suppliers that support each manufacturing business and customer. These businesses provide services and products as well as maintenance spares and often they are forgotten elements of the sustainability puzzle. Many businesses do not extend their policy deployment systems to suppliers or forget to invite suppliers to engage with improvements. Involving suppliers makes perfect sense – they are experts and they are dependent, to a greater or lesser extent, on your business. So it makes sense to involve them. Many suppliers will happily engage in Kanban extensions or use vendor-managed inventory procedures to allow customer businesses to have a full availability of stocks. Suppliers are also important sources of cost-reduction ideas including lightweighting products, reducing variety, introducing product traceability systems and finding ways of being more environmentally compliant.

There are two mechanisms for engaging suppliers, the first is on a one-to-one project basis and the other is to create a collective of suppliers to serve as a focused improvement network. The latter is a typical form used by Japanese world class businesses. The network is convened at regular times during a year. At meetings the business strategy of the customer is shared (so that suppliers know and can make better decisions to support the customer) and methods are exchanged/shared so that all suppliers can benefit. So if your business demands the introduction of barcoding or a new ISO quality management system then it makes sense to share risks and do it as a group. That way the entire supplier chain can determine which standard of barcode to use and how it will be used to enable better information flow between businesses. Creating this external team is very important and a natural extension of the policy deployment and Lean TPM systems.

9.5 CHAPTER SUMMARY

There is no such thing as a 'recipe for sustainability' in terms of a solution that fits all businesses. Even companies making the same products with identical technology will differ in terms of their ability to sustain (anyone who has experienced work in a large divisionalised organisation will appreciate this statement). The recipe is individual and specific to the firm – it is also time-specific and therefore will change especially during conditions where new senior managers move into the firm (this type of strategic management rotation is also a means of preventing sustainability and creating a lack of strategic direction). So companies must decide for themselves what fits and what does not. To some the idea of a common canteen shared by management and workers will be seen as taking things too far – that speaks volumes about the current culture. Sustainability and culture change are not the results of ticking a few boxes on a 'pick and mix' menu of good practices. It takes time but more importantly it takes planning. Rational behaviour and improvement activity must be nurtured, discretions should be identified and highlighted and persistent employees who actively attempt to derail the programme of change must be dealt with. If they are not then that will become a precedent of defiance and reason why people will not take part in such improvement work. Obviously there is a trade-off in terms of actively seeking questions from employees and seeing these as attempts to subvert improvement. Even some of the brightest people have pretty poor interpersonal skills which to others will come across as offensive or intimidating. That is OK we are all human but that does not mean these behaviours should not be pointed out to the individual and action taken to address these when they become persistent offences and cause insult. Such events must be handled on a 'case-by-case' basis but when they compromise the basic principles of the firm, the individual must face a decision to comply with the wishes of the workforce or move on. That is the hard side of sustainability – but it has to be faced. You cannot build a 'world class' business with people who deliberately and persistently defy 'reasonable requests' to change. It is also not fair for the individual.

Having said all this, most employees, even hardened cynics, are capable of sustaining change and most employees enjoy the new skills and variety of working life that Lean TPM brings with it. Further, most would not want to go back to the 'good old days' – because they simply were not good. At the end of the day the mortgages and holidays of every employee rest upon the performance of the firm, so planning for sustainability and autonomous improvements is mandatory. Culture will change as a result – but this happens much more slowly.

REFERENCES

Belcher, J. (1987). *Productivity plus: How today's best companies are gaining the competitive edge*. New York: Gulf Publishing.

Goldratt, E. (1984). *The goal*. Aldershot: Gower Publishing.

Kohn, A. (1999). *Punished by rewards*. New York: Houghton Mifflin Publishing.

Liker, J., & Meier, D. (2008). *The Toyota way fieldbook*. New York: McGraw Hill.

Peters, T. (1992). *Liberation management: Necessary disorganization for the nanosecond nineties*. London: MacMillan Publishing.

Sobek, D. K., & Smalley, A. (2008). *Understanding A3 thinking*. New York: CRC Publishing.

Standard, C., & Davis, D. (1999). *Running today's factory*. Cincinnati: Hanser Gardner Publications.

Index

Note: Page numbers followed by "f" and "t" indicate figures and tables respectively.

Printed in the United States
By Bookmasters